Ocooch Mountain Rocks

Scary Rock Found by Grandson

Ocooch Mountain Rocks

"Ancient Cave Stone"

Spiritbooks LLC

Eric Jonathan Zingler

Library of Congress Control Number: 2024911857

Copyright © 2024 Eric Jonathan Zingler.

All rights reserved. No part of this publication may be reproduced,
distributed, or transmitted in any form or by any means, including photocopying,
recording, or other electronic or mechanical methods without prior written permission
of the publisher or author, except for brief quotations in book reviews.
For permission requests, contact the publisher.

Pictures presented in the book were taken by Eric Jonathan Zingler,
with all rights reserved unless attributed to another photographer or website.
No picture dates are given next to Zingler's photos. Features were captured spontaneously
at different times with a camera or cell phone. They are unique images.

1st Edition, 2024

Spiritbooks LLC
24403 Lyndale Rd
Kendall, Wisconsin 54638
www.spiritbooks.xyz

Printed in the United States of America.

eBook ISBN: 978-1-964073-00-2
Paperback ISBN: 978-1-964073-01-9
Hardcover ISBN: 978-1-964073-02-6

The book cover photo is courtesy of Joy Nangle, 2021.
Interior title page image, Dave Bunnell, *Exploring a sea cave in Mendocino County, California*, December 30, 2005, CC 2.5, https://commons.wikimedia.org/wiki/File:Seacaving.jpg. The image was chosen for the lack of plant life which was not present in the Ordovician. The human figure was removed since the scene represents a cave formation near a shallow, warm sea 480 million years ago.

To my sister Bonnee and her husband's cousin who gave me Ocooch Mountain Rocks from their land, others who allowed me to obtain exceptional specimens, people who came to see my displays in the yard, especially the little girl who visited twice and brought her rocks along, those who encouraged me to write about them, farmer friends who hauled the stones to our acreage, the Amish farmer who gave me a river gravel, son Jonathan for help with publishing, and my cat, BC, who took me on walks, climbed trees and explored many rock piles.

Ah, you have a familiar looking collection of bizarre rocks, which have had an amazing diagenetic history that almost defies a full understanding-

Robert H. Dott Jr., PhD, University of Wisconsin-Madison

Contents

Origins   1

Rock & Life   13

A Gallery of Strange, Beautiful, & Scary Rocks!   19

Bibliography   86

Index   88

Preface

My mother told many stories about my great-aunt Tina, who was an artist. As a boy, I drew birds, but college, a big family, my wife and I had eight children, and work suppressed the artist within me. My effort through a companion book, *Ocooch Mountains,* to discover the origin of Ocooch started a writing career. Initially, the *Ocooch Mountain Rocks* were to be a chapter in the book. After learning about these rocks' unusual and enigmatic origins, I realized a second manuscript was necessary.

Rocks fascinate people, especially exotic ones, and the Ocooch Mountain Rocks give an exceptional, over-the-top visual experience that is visceral and emotive. Several times, we held an open house where people toured my displays, and no one was disappointed. Many returned for another look. One little girl mentioned in the dedication came twice and, on the second trip, brought her rock collection to share. That was a great day! The book was written because people wanted it and encouraged me to write.

I love history, particularly earth history, which was my favorite part of the encyclopedias growing up as a latchkey kid in a small town. Given the highly unusual nature of the rocks, petrified structures formed by an early lifeform, the book became an opportunity to present a unique view of geological history for our area that covers billions of years, tracing the lifeform through time and planetary changes to our present-day, explaining bestowed benefits, and how the rocks were formed. It is a story about the symbiosis of rock and life.

I have a gift for seeing patterns and catching glimpses of unusual shapes when walking in my rock gardens or seeing a pair of eyes looking at me when sitting in the firelight surrounded by a jagged, horse-shoe-shaped wall of rocks. Sometimes, an unusual and surprising visage appears when turning a rock on bluffs in piles of stones placed there by countless hours of toil as farmers removed them from the field where they weathered up and posed a danger to equipment. Some features of interest may appear after rain or when power spraying to remove dirt and excessive lichen and moss. Magnification can also uncover startling images. In the magic moment of discovery, pictures are spontaneously taken with whatever cellphone or camera is available.

As explained in the book, the rocks are formed by complex interactions with water, which brings out the rocks' unique colors. Adjusting for light and contrast often reveals hidden features and a variety of minerals and quartz crystals. They are one-of-a-kind photos, even though some patterns repeat. All capture imagination and evoke emotion. Realizing their extraordinary visual impact, a gallery of exceptional images is presented that transformed the manuscript into more than a geology book; it is photographic natural art, a story in stone.

Another reason for the book was to leave a legacy for great-grandsons and daughters so they would have more than a picture in an album to recall their grandfather. The book reflects curiosity about the outdoors, nature, and the beauty found even in rocks. Hopefully, the book inspires a desire to look beyond the surface of things to discover something mysterious and beautiful. Art positively changes those who experience it by giving a different perspective. If rocks can be transformative, this effort may encourage finding hidden beauty within people. Both books fueled a passion for writing unique history books, and more manuscripts are being written. So, I have become an artist and can take my place with Aunt Tina in the family history!

I want to acknowledge Dave Lovelace, PhD, University of Wisconsin Geology Museum Scientist, for comments on composition, formation, and review of the book; Joanne Kluessendorf, PhD, for composition opinion; and Robert H. Dott Jr., PhD, for his published research and encouragement. If you like rocks, you will love this book, and if you have no prior interest in geology, it may awaken a new passion for knowing more about our home on Earth.

Eric Jonathan Zingler, 2024

The image shows dense forests releasing organic compounds that create a hazy look accentuated by mist. Forested peaks, bluffs, and ridges make Ocooch Mountain Rocks hard to find.

Dandog77, *Wildcat Mountain*, September 9th, 2008, CC 3.0 https://commons.wikimedia.org/wiki/File:TypicalDriftless.jpg.

Origins

I grew up in a small village surrounded by peaks, bluffs, and ridges. My two favorite places to play were by a creek and a marshy pond behind our house. Pollywogs were collected each spring. When I caught fish, I tried to keep them in my mother's ceramic barrel. I also brought home frogs in coffee cans, clams for the fish bowl, and crayfish. Once, I got a big bunch of crayfish and put them in my mother's washtub. The next day, they were gone! I asked my mother where they were, and she said, "They got away." I never could figure out how they scaled those tall walls! Later, I deduced that my dad had a role in their escape. Hints came when my mother sometimes told a story that made her laugh about a neighbor calling out to my dad when he was swinging his ax to chop a hole in the creek ice. The neighbor questioned his sanity. He answered that he had to "get the frog out of the house!" So, I suspected he took care of the crayfish, too. His worst job from overheard conversations was when I used my mother's new white pail to catch a nice long black bloodsucker. He had to use that trusty ax again to chop up the pail to appease her! His ax hangs on my garage wall next to her washtub.

Another creek flowed under a bridge south of where we lived. A large pool had formed on the east side. Occasionally, you could see a large snapping turtle rise near the edge and disappear into the dark depths. If you were lucky, you might see it walk out on land to lay eggs. The water was not so deep under the bridge and flowed very fast. Schools of iridescent fish the size of guppies would float in the oxygenated stream of water, and red-orange crayfish could be seen. They were too beautiful to catch. In the shallow, clear water, I could see brightly colored rock, and I added some to my collection, my first encounter with the Ocooch Mountain Rocks.

My mother had a flower garden in the back of the house next to the lilac bush where the hill sloped. She arranged it like a terrace with nice, rich soil behind stacked rocks. After my dad passed away when I was ten, I did my best to help her weed it. The stones were interesting, but one stood out, and I wondered where it had been found. It was very different from the others, and when I asked my mother, she said it came from my sister's farm. My sister lived about three miles from our home at the end of a long driveway just below the top of a tall bluff where you could see for miles.

After I went to college, married, and started a family in another state, we would often come home to visit my mother. All the kids, we had eight, loved to visit Grandma, who did everything possible to make it a happy event. Everyone had their cot to sleep on, and the food was terrific. Grandma was the best cook ever!

A young couple lived in a big house next to Grandma that seemed super tall, probably from its placement on the upper side of the small hill where we lived by an elm, the tallest tree in town. In grade school, my hometown was a forest of elm trees lining the streets on either side. I could ride my bike all the way home under their shade. After they died off, maple trees were planted to replace them. As time passed, I noticed rocks would appear in the neighbor's yard. Some were spectacular, with holes and all sorts of twists and strange shapes. Then, a rock garden with a fishpond could be seen, and I became jealous! I finally had to ask the neighbor where they were getting these rocks. He replied that his wife found them on a hillside where water flowed down a gully. This was a complete mystery to me. Driving through the area, I could only see blah, washed-out gray weathered sandstone. I had never seen anything like their rocks anywhere!

When we moved into a bigger home with a nice backyard, I daydreamed of a pond like Grandma's neighbor. No rocks were comparable in our city, so on a visit to Grandma, I asked my sister where Mother got that big, strange rock. She told me to check out the old gravel pit dug for sand and gravel when cars replaced horses and better roads were needed. It was small, only a few hundred feet long, and about ten to fifteen feet deep. A companion pit dug about the same time on a bluff across a deep, wide valley from my sister's farm was even smaller. The location of the smaller pit seemed odd. It was almost a quarter mile across the field from the curvy road that snaked up the hill. Why didn't they dig out sand and gravel at a more convenient place?

I was surprised and pleased to find several large and strangely formed rocks in my sister's gravel pit near the edge, like the one in my mother's flower bed. Smaller, colorful stones could be seen in a cave-like area covered with brush and dirt. With the help of my son, several carloads of rocks were transported home each time we visited Grandma. Finally, a pond was built. Soon, a water lily grew, and several frogs from a nearby marsh found a home in the pond.

More time passed, and eventually, we moved back to Wisconsin to live on a hill that was once part of the family farm across the deep valley from my sister's farm and next to the mesa-like bluff with the smaller gravel pit. Of course, I brought most of my rocks with me! A farmer friend who saw the stones told me he knew of a place where there were a lot of strange rocks like mine. A cousin of my brother-in-law owned it. His land joined my sister's farm, and his rocks were amazing. There were several piles. Some were so striking that he placed them near his wife's flower garden.

I asked if I could have some, and he said yes, there was a large pile on the edge of a ridge by a field to the west. He also told a story about drillers who came with a rig in the 1940s looking for sand and gravel. The shaft and bit disappeared when they started drilling in the field! Scared that the rig's weight might cause a cave collapse, they hurried out of there and later found a small area of suitable sandstone on my sister's farm at the northwest edge of her bluff, which became the gravel pit where I found my pond rocks.

The way to the ridge took me up a gully that reminded me of a crystal staircase. Block-like stones with quartz sparkled at intervals up the slope that running water had exposed. I thought of my mother's neighbor and how she found rocks in a similar place. I could see colorful rocks scattered and an extensive pile at the top. Farmers had picked rocks that weathered up in the field and placed them at the edge for decades. There were so many different shapes and sizes that I could not choose what to take, so my farmer friend used his tractor shovel to scoop up a pile of rocks and place them on a wagon. He took a couple of loads to my place, and I spent months sorting through the stones and arranging them in displays. I wondered if the tan-colored rocks indicated a reef.

While researching material for my book, *Ocooch Mountains*, I discovered that Major Stephen Harriman Long first described rocks similar to my collection. He wrote about them in his journal entitled *Voyage in a six oared skiff to the Falls of St. Anthony in 1817* when he charted the upper Mississippi River:

> On the top of the hill, we collected many exciting specimens of minerals, such as crystals of iron ore, silicious crystallizations beautifully tinged with iron, some of them purple, others reddish, yellow, white, etc., crusts of sandstone firmly cemented with iron, and I think set with solid crystals of quartz, etc.[1]

In her seminal work *Wisconsin's Foundations*, Gwen Schultz associated these rocks with the Prairie du Chien dolomite. She described the rocks as generally "light buff, gray, off white, and sometimes white" that contained "flint or chert nodules" with some "brecciated" rock, which is a term for angular broken pieces rolled together by waves near the shore and re-cemented. Cavities in the dolomite were distinguished by crystals that filled smaller ones and lined larger. Some cavities grew to cave size. She quotes T. C. Chamberlin, a University of Wisconsin geologist who described the crystals:

> Where the cavities are larger, the crystals only form a lining, producing drusy little grottoes, some of which are very beautiful. The quartz is most frequently transparent or opalescent, but it is sometimes red, brown, or rose colored. The crystals are sometimes grounded on a chalcedonic base, forming a beautiful combination.[2]

In his classic work *The Physical Geography of Wisconsin*, Lawrence Martin credits Chamberlin for explaining how the Driftless Area of southern Wisconsin escaped glaciation, preserving the dolomite.[3] Schultz said the thickness of the Prairie du Chien dolomite varied due to extensive erosion when an epi-continental sea withdrew. Rivers cut deep valleys that reached the Oneota sandstone of the lower Ordovician Period approximately 480 million years before the present (BP) and still deeper to the earliest Cambrian sandstone called Mount Simeon. Chamberlin called the landscape "petrous billows." Arid conditions, perhaps a desert, created strong winds that whipped up the earlier Cambrian sands and ground them to fine granular shapes. Deposited over the wavy Prairie du Chien formation, the St. Peter sandstone varies from quite deep to completely absent. Schultz says the sea returned, and beach and lagoon remnants are present with "complex undulations, ridges and dunes; some are more than 30 feet high."[4]

It was clear that I had a mystery to solve. Were my rocks those described by Major Long, Schultz, and Chamberlin? I emailed photos to the geology department at the University of Wisconsin, my Alma Mater. Dave Lovelace, PhD, Museum Scientist, University of Wisconsin Geology Museum, replied that I had not found a reef as I might have imagined but rather sedimentary layers of stromatolites. The name is derived from the Greek word strōma (layer). His research indicated that the rocks were most likely associated with the Oneota sedimentary sandstone of the lower Ordovician, part of the Prairie du Chien dolomite. He thought the area might have been an offshore shoal where stromatolites accumulated to reef height. He explained that the strange shapes were created over time: "This wasn't really caused by any one event, but a process called diagenesis [chemical alteration]. The colors come from varying concentrations of minerals such as iron, manganese, magnesium that are bound in the otherwise relatively clear crystalline quartz. Formed just as Chamberlin described."[5]

---

[1] Major Stephen Long, *Voyage in a six-oared skiff to the Falls of St. Anthony in 1817* (Philadelphia: Henry B. Ashmead, Book and Job Printer, 1860), 49-50.
[2] Gwen Schultz, *Wisconsin's Foundations* 2nd ed. (Madison: University of Wisconsin Press, 2004), 90.
[3] Lawrence Martin, *The Physical Geography of Wisconsin* (Madison: University of Wisconsin Press, 1965), 119.
[4] Schultz, *Wisconsin's Foundations*, 90
[5] Dave Lovelace, email to Eric Zingler, October 23, 2012

The sedimentary lines in the Ordovician rock quickly identified the whitish rocks in my benefactor's rock pile and garden. They were all stromatolites ranging from small circular ones to large round specimens that experienced extreme weathering. I also emailed Joanne Kluessendorf, PhD, Director of the Weis Earth Science Museum, University of Wisconsin-Fox Valley at Menasha, and attached photos. She confirmed Lovelace's response. My rocks were not part of a reef as we think of one. Nevertheless, an amazing photo caught her eye! It looked like coral! However, she decided the structure was most likely small stromatolite cavities. Another picture intrigued her. She wondered if part of a trilobite fossil was near the center.[6] Trilobites were marine animals that proliferated in the Cambrian Period, which preceded the Ordovician.

Robert H. Dott Jr. and John W. Attig's book *Roadside Geology of Wisconsin* also confirmed that the area where I live had been covered by a shallow sea, perhaps only twenty or thirty feet deep. They cite a well-known geologist, Laurence Louis Sloss, who bragged: "The early Ordovician Sea was so shallow I could have walked through it." Dott Jr. and Attig wrote that photosynthetic bacteria thrived in warm, salty environments, which hindered predatory organisms. The bacteria secreted a mucus coating that trapped sand from wind or tidal currents, which blocked sunlight, so the bacteria had to secrete a new layer of mucus on top of the previous layer. Endless repetitions of this cycle created sedimentary layered structures called stromatolites, identified in their chemically altered form by Lovelace and Kluessendorf from the pictures I sent. Stromatolites varied in size from "small biscuit-like surfaces to large turban-shaped domes."[7]

As Lovelace explained, stromatolites formed in tropical offshore shoals. Schultz described how tides broke up structures; some small pieces were washed around, accumulating precipitated calcite. These rocks appear as tiny round ball bearings called oolite, a sign of shallow tide-agitated water.[8] Modern-day stromatolites are only found today where super salty water protects them from predators.[9] Water movement influences the stromatolite structure.[10] Tops become round as they orient to the sun and may have stems in deeper water.

Oolite, Paul Harrison, *Stromatolites,* Hamelin Pool Marine Nature Reserve, Shark Bay Australia, 2005, CC 3.0, File:Stromatolites in Sharkbay.jpg, *Shark Bay Australia,* Mindat, Hudson Institute mindat.org/photo-595501.html

---

[6] Joanne Kluessendorf, email to Eric Zingler, November 7, 2012
[7] Robert H. Dott, Jr. and John W. Attig, *Roadside Geology of Wisconsin* (Missoula: Mountain Press Publishing Co., 2004), 16, 198.
[8] Schultz, *Wisconsin's Foundations,* 195
[9] David L. Alles, ed., "Stromatolites," Western Washington University, PDF, 3, last updated March 28, 2012, https://fire.biol.wwu.edu//trent/alles/Stromatolites.pdf.
[10] "Stromatolites," Sharkbay, accessed December 15, 2020, https://www.sharkbay.org/publications/fact-sheets-guides/stromatolites/.

While sorting through my rocks, I discovered an incredible find: an eroded and silica-transformed intact stromatolite. How it survived for 480 million years is a mystery. The next photo shows significant alteration by diagenesis and breakage, although sedimentary lines are still visible in a dome-like fashion. The following image is another rare discovery of a stromatolite stem. The dome might have collapsed inward.

Eager to learn more about the stromatolite fossils that experienced diagenesis in our area, I searched the Internet for academic papers. I found an article, "Authigenic Silica Fabrics Associated with Cambro-Ordovician Unconformities in the Upper Midwest," by George L. Smith, Robert H. Dott Jr., and Charles W. Byers, published in *Geoscience Wisconsin*. Despite the daunting title, they were writing about my rocks!

Authigenic means formed in the place, *in situ*, rather than transported from another location. Silica is quartz. Fabric refers to the elements and geometric configuration of rocks. Breaks in sediment formation are called unconformities since the rock above is different. However, the paper stated that pieces of these rocks could be found above the gap, indicating they may have formed before or when erosion created the sedimentation break.

The journal article of Smith, Dott Jr., and Byers stated that limestone sediments are commonly silicified in a dense, finely-grained, homogeneous manner. In contrast, limestone associated with unconformities in the Mississippi Valley region, where I live, was "markedly different" with cavities referred to as "very vuggy (with drusy quartz linings)" as Chamberlin described, "botryoidal" (like a bunch of grapes), heterogeneous, and in many cases, mega quartz crystals formed.[11]

Amethyst and large (mega) quartz crystals on stromatolite stone, my sister's keepsake. Her son found the gem in the gravel pit on their farm.

---

[11] George L. Smith, Robert H. Dott, Jr., and Charles W. Byers, "Authigenic Silica Fabrics Associated with Cambro-Ordovician Unconformities in the Upper Midwest," *Geoscience Wisconsin*, 16 (1997): 25, 28.

I realized Dott Jr. was *Dr. Dott*, my instructor when I took a geology course at the University of Wisconsin. He and his co-authors wrote that the origin of these rocks had been a mystery. Their location was also peculiar. The stones were "patchily distributed" under the unconformities with varied degrees of silicification, a term for silica or quartz replacement. Iron oxides and hydroxides account for vivid red, green, yellow, blue, purple, brown, and black colors. Their article's map of well cuttings shows a significant presence of these rocks in the Driftless Area.[12] The Driftless Area, known as the Paleozoic Plateau, eroded into peaks, bluffs, ridges, and deep valleys. The geologic term Driftless means no glaciers plowed through and erased landforms. The Driftless Area includes southwest Wisconsin and parts of northwest Illinois, northeast Iowa, and southeastern Minnesota. The Mississippi River flows through the area in a north-south direction. The Wisconsin River, a major tributary

of the Mississippi, borders the Wisconsin portion of the plateau on the east. It cuts a broad valley in a southwest course toward Prairie du Chien, where it joins the Mississippi. The largest segment of the Driftless Area in southwestern Wisconsin is the Ocooch Mountains. The northern edge of the Ocooch Mountains is seen as a blue wall in the adjacent photo. I live on the jagged edge where a hill obscures the wall.[13]

Bryon Bootman, *Paleozoic Plateau,* 2019.

Dr. Dott and Attig explain why Ordovician rock is present:

There is practically no rock record for a 400 million-year interval from the Devonian to the Quaternary Ice Age. The only possible deposits for the missing interval occur in western and central Wisconsin as widely scattered ridgetop remnants of very poorly dated river gravels. Although geologists have assigned them to the Cretaceous period, they could have been formed anytime within the last 130 million years. So we can only wonder which dinosaurs wandered over Wisconsin and which ancestors of woolly mammoths grazed here during that long gap in our geological record.[14]

Erosion that reached the lower Ordovician is related to the rising of the Wisconsin Dome, or as Schultz mused, the surrounding area "sank."[15] The increase in elevation enabled rain, streams, and rivers to erode the geologic time periods away. Martin says that one ancient river bed runs parallel to the west of the upper

Kickapoo River, and he estimated the height of the ridge tops "have been lowered at least 800 feet." The gravels are composed of rock that resists erosion, such as chert, and well-rounded shapes indicate great length.[16]

I discovered one of these rare river gravels! On a sunny summer afternoon, I drove west along the edge of the blue wall in the photo, continued past the area where the springs of the Kickapoo flow, turned south, and stopped when I saw an Amish man with a team of horses and a wagon coming up the road. I asked him if he had ever seen a smooth, round rock on his land. He answered, "Sure enough," and said to follow him up the road to his place. We walked over to a shed where he showed me stones that he said looked like "a pile of potatoes." My eye caught a particularly round rock. I told him I collected rocks and asked if I could have the round one, and he generously agreed. It stretches the imagination to stand where these river gravels are found and visualize a floodplain with slopes that tower 800 feet! Other remnants of these ancient rivers can be found at Seneca, Crawford County, Windrow Bluff in Monroe County, and the Baraboo Range near Devil's Lake State Park.[17]

The waxing and waning of the salty, shallow Ordovician epi-continental sea created the unconformities Dr. Dott and his colleagues mentioned in their academic paper. Sediments that accumulated in the shallow sea are

---

[12] Smith, Dott, Jr., and Byers, "Authigenic Silica Fabrics," 25, 26, 28, 30
[13] Eric Jonathan Zingler, *Ocooch Mountains* (Kendall, WI: Spiritbooks LLC, 2024), 1.
[14] Dott, Jr. and Attig, *Roadside Geology of Wisconsin,* 20-21
[15] Schultz, *Wisconsin's Foundations,* 13
[16] Martin, *The Physical Geography of Wisconsin,* 53
[17] Martin, *The Physical Geography of Wisconsin,* 53

called the Oneota. As it retreated and then returned, the Shakopee was deposited. As Schultz explained, the Ordovician sedimentary deposits are collectively called the Prairie du Chien dolomite, forming the capstone of the Driftless Area peaks, bluffs, and ridges. Dr. Dott and his co-authors say the intense silicification of stromatolite rock with "large chalcedony-and quartz-lined vugs" is unique in the Prairie du Chien dolomite. They say "tubular vugs," the "molds of vertically stacked stromatolites," are distinctive in the Oneota strata.[18] This confirmed Lovelace's research and explained why a rich trove of colorful, strange-shaped rocks can be found where I live. The elevation of our acreage is at the capstone level.

Dolomite is limestone mixed with magnesium, a vital mineral for life. Magnesium makes rocks more resistant to weathering and gives a castellated look to many peaks and bluffs. As the land was exposed to rain, streams, and rivers, water interacted with the dolomite to create carbonic acid that carved cavities in the limestone. Chemical interactions created crystals, Chamberlin's "drusy little grottoes."

The association of Ocooch Mountain Rocks with unconformities in a "patchily distributed" pattern with varying degrees of silicification accounted for my experience of miles of sandstone outcroppings with no unusual properties. High-quality Ocooch Mountain Rocks are hard to find! "Intense silicification" explains the extraordinary specimens and colors.

Carbonic acid and cavities are associated with cave formation. Sorting through my piles of rocks revealed what appeared to be broken pieces of stalagmites and stalactites. Lumpy pieces where water hit the bottom may be stalagmites, and sharper points where water dripped from the ceiling could be stalactites. One long piece featured in the picture behind a lumpy fragment has a pebble fused to its tip, presumably cemented in the rubble after a cave collapsed.

A tan rock in the cave fragments may be a piece of cave ceiling. The dark streak on its surface might be mineral accumulation by dripping water. The exterior of these stones appears dull and undistinguished. However, the stone with a pebble fused to its tip broke and revealed an incredible world of color. A pointed columnar piece seems to be an exceptionally chemically altered stalactite.

The property of my brother-in-law's cousin is in a rounded funnel-shaped valley with an open end adjacent to the bluff my sister owned. The valley rock piles appear to indicate a cave collapsed, and water washed out broken pieces of rich specimens as the valley eroded over millions of years. The extraordinary rocks at the edge of the ridge were picked as they weathered up in the bluff field from a layer of stromatolite stone or as cave fragments. This rich area may indicate an intact cave still exists, which would give credence to the story of the drillers losing their bit and shaft and quickly moving on to where they found the small amount of sand and gravel on my sister's land.

Unusual rocks were found on both ends of my sister's bluff. I also discovered extraordinary specimens in rock piles on the high mesa-like bluff near my home, across the deep and wide valley that divides the bluff from my sister's property. The rich lode of Ocooch Mountain Rocks on this ridge remnant explains the odd placement of the smaller gravel pit, which appears to have been an area of sandstone surrounded by stromatolite rock at reef heights. After reading Dr. Dott and his colleagues' research paper, I wrote an email to him, attached pictures, and asked for his opinion on what he saw:

---

[18] Smith, Dott, Jr., and Byers, "Authigenic Silica Fabrics," 31

Dear Eric,

Ah, you have a familiar looking collection of bizarre rocks, which have had an amazing diagenetic history that almost defies a full understanding. From the variety of textures, etc. that I see plus your notation of elevation around 1300 feet, I feel sure that these are all blocks of much altered (by diagenesis) remnants of the Prairie du Chien Dolomite formation. This formation caps most of the hills in western Wisconsin that are of the order of 1300 feet or so. You will find it discussed in our Roadside Geology book. The dolomite, which is composed of calcium, magnesium and carbonate (CO3), is fairly soluble in groundwater that is slightly acidic. This leaches the dolomite to produce a lot of cavities of size varying from an inch or so to cave size. That has happened to the PdC dolomites all over the region and probably the leaching has happened over much of geologic time since the rock first was deposited about 480 million years ago. An important part of the chemical changes in this rock (diagenesis again) involved silicification, that is the chemical replacement of dolomite by silica (= chert and quartz). I see a lot of evidence of this silicification of your boulders. In fact, I suspect that 90+ percent of the original dolomite has been replaced by one form or another of silica. Several of the photos also suggest a lot of iron concentration, too. I think that much of all this alteration occurred during the exposure of the formation to the atmosphere before the deposition of the non-marine St. Peter Sandstone, which covered a widespread unconformity that developed during the droop of sea level, which affected the entire Midwest region of the country. The St. Peter was deposited by wind and running water (i. e. rivers). Soil-forming processes caused much of the alteration you see combined with groundwater flow through the rock, which dissolved out the caves, some of which are popular today for spelunking. But cave formation doubtless has continued later, probably up to the present time. Much of the silicification probably occurred within the upper soil zone. Such a process is seen in modern soils in several parts of the world.

I cannot explain every detail of these kinds of altered rocks. In some cases, it appears that fossils, especially the stromatolites, have been preferred sites for silicification. Why? I don't know, but obviously silica-bearing solutions could react with them most easily. Because all the fragments are angular, we call this a breccia. They are common in the Prairie du Chien but are randomly scattered rather than forming a stratum. In fact, most of them seem to have formed long after deposition in areas undermined by solution, as in a cave, and then the 'roof' collapsed to form the broken, angular rubble called breccia. Breccia also can form along faults, but I doubt this is the case for yours.

So you have a jumble of mostly chert, which has replaced dolomite. Stromatolite remnants are still recognizable, but other original textures and features have been virtually obliterated by that nasty diagenesis. There is a ripe field for some hardy geologist to tackle a detailed study of the diagenesis of the Prairie du Chien dolomites. I wish…someone with the right kit of tools…and the incentive…[would] tackle this tough problem. Our former student, George Smith, lead author of that article you found, just scratched the surface. I hope this helps you a little.

R. H. Dott, Jr.[19]

Dr. Dott passed away in 2018. Joanne Bourgeois wrote a tribute to him in *GeoScienceWorld*. At the University of Wisconsin-Madison, he "educated and inspired thousands of students of earth history, built an eminent research program in sedimentology and stratigraphy, and carried out distinguished scholarship in the history of geology."[20] The following images have comments from Dr. Dott, Lovelace, and Kluessendorf.

---

[19] Robert H. Dott, Jr., email to Eric Zingler, December 12, 2012
[20] Joanne Bourgeois, ROBERT H. DOTT, JR. (1929–2018), EARTH HISTORIAN AND HISTORIAN OF GEOLOGY, *GeoScienceWorld*, 38, 2 (October 1, 2019), 1, https://doi.org/10.17704/1944-6178-38.2.422.

"[R]eally bizarre... ropy rock now replaced with silica," Dr. Dott.

"Hemispherical stromatolite," Dr. Dott,
"manganese," mysterious black band, Lovelace,
oncolites, stromatolites in a raised circular form, stromatolites in
circular biscuit form, PaAt-56, *Crayback Stomatollite*, Nettle Cave,
Jenolan Caves, Australia, CC 3.0, similar ridges are on the inset.
Donnie Reed, *Microbialite towers*, NASA, inset a fossil tower?

**Circular Dome
May have been collapsing.**

Kluessendorf, "iron crust," sedimentary lines are still visible. Stromatolite cavities filled with orange, yellow, reddish, and purple-colored quartz, one resembles brain coral. However, Lovelace said brain coral did not evolve for almost 100 million years later!

Rock & Life

The photosynthetic bacteria referred to in *Roadside Geology of Wisconsin* are called cyanobacteria. The link between cyanobacteria and the Ocooch Mountain Rocks is almost as old as the earth. Cyanobacteria evolved shortly after the craton under the Ocooch Mountains formed as one of the first crustal pieces. When you pick up an Ocooch Mountain Rock, you hold mineralized fossils of early life that traveled through time with the deep roots of the land you stand on. The story of their journey adds a sense of wonder when the Ocooch Mountain Rocks are found.

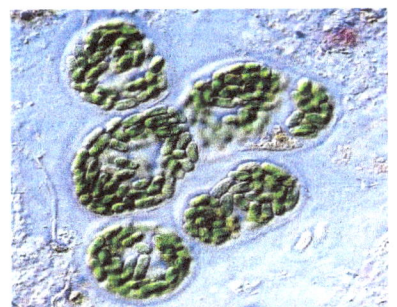

Unicellular bacteria from a microbial mat in Guerrero Negro, Baja California, Mexico, no date given, NASA, File:Cyanobacteria_guerrero_negro.jpg

Over four billion years ago, crustal pieces began to form on the earth. Eventually, the Superior Province, which implies uniformity, accreted in the water world of Earth at that time. James Palacas, in his work for the US Geological Survey, writes that the Superior Province covers areas of Minnesota and northern Michigan with a section underneath western counties in Wisconsin.[21] The Superior Province was part of the heart of the Canadian Shield, also called the continent Laurentia, which became North America.

The moon was much closer as the sun rose in a hazy sky shortly after the Superior Province formed. Great tides swept over early land, washing mineral nutrients into the sea. A miracle happened; life appeared. During crustal formation, acid rain and plate tectonics destroyed early landforms, so evidence was hard to find. Tia Ghose wrote an article for *Live Science* and quotes Robert Hazen, an earth scientist, about the discovery of fossil cyanobacteria: "This is one of the, or the, oldest fossils ever found. You've got a 3.5-billion-year-old ecosystem." Ghose describes how Hazen's colleague, Nora Noffke, found the fossils in the Jack Hills of western Australia, part of the ancient Yilgarn Craton: "So many geologists have walked over the same rocks and never noticed anything…Under a microscope, the formations revealed a series of individual black filaments intertwined with sand grains characteristic of microbial mat communities."[22]

Tim Bertelink, *Archean Eon*, March 21st, 2017, stromatolites are in the water, CC 4.0, File:Archean.png - Wikimedia Commons.

The research of Abigail Allwood and colleagues published in *Nature* on the approximately 3.5 billion-year-old Strelley Pool Chert located in one of the earliest earth cratons, the Pilbara Craton in northwest Australia, revealed seven different kinds of cyanobacteria had evolved.[23] Equally impressive, the research of Tara Djokic and co-authors of a study published in *Nature Communications* on cyanobacteria fossils in the Pilbara Craton Dresser Formation, as old as crustal pieces of Mars, shows cyanobacteria were thriving in hot springs.[24] Another exciting discovery in Greenland by Allen P. Nutman and fellow researchers published in *Nature* predates Pilbara. They have found evidence of stromatolite presence dated to 3.7-3.8 billion years BP in

---

[21] James Palacas, "Superior Province (051)," USGS, Energy Data Finder, PDF, accessed July 15, 2019, https://certmapper.cr.usgs.gov/data/noga95/prov51/text/prov51.pdf.
[22] Tia Ghose, "3.5-Billion-Year-Old Fossil Microbial Community Found," *Live Science,* November 13, 2013, https://www.livescience.com/41191-ancient-microbe-fossils-found.html.
[23] Abigail Allwood et al., "Stromatolite reef from the Early Archaean era of Australia," *Nature,* 441 (July 2006): abstract, DOI: .1038/nature04764.
[24] Tara Djokic et al., "Earliest signs of life on land preserved in ca. 3.5 Ga hot spring deposits," *Nature Communications,* 8, 15263 (2017): 7, https://doi.org/10.1038/ncomms15263.

southwest Greenland. Trace elements indicate a warm, shallow sea. At the same time, Mars had water.[25] These discoveries underlay a fascinating paper by Vincenzo Rizzo and Nicola Cantasano in the *International Journal of Astrobiology* that makes a case for stromatolite formations on Mars from pictures taken by NASA rovers.[26]

After its emergence, the Superior Province and the cyanobacteria began a long journey through the supercontinent cycle of land that broke up and then came together again as, metaphorically, the earth breathed. Supercontinent refers to a time when major land terranes came together as a unified mass. The first pieces that collectively came together are hypothesized to be Vaalbara, although very small. The name is the last four letters of two ancient cratons, the Kaapvaal Craton in southern Africa (Vaal) and the Pilbara Craton in Australia (bara). The geological features are very similar. In their study published in *Precambrian Research*, Anil Kumar and colleagues show a paleomagnetism and geochronology link between these two cratons and the India craton Singhbhum, suggesting ancestry in Vaalbara.[27] After Vaalbara, Ur formed. S. Mahapatro and researchers say in their *Geological Journal* article that the largest preserved segment of Ur is in the Indian Shield.[28] Crust grew around Ur, then splintered as, metaphorically, the earth exhaled.

The Algoman or Kenoran Orogeny is the first identified mountain-building event in the Superior Province area. Research by Charles Gower and Paul Clifford in the *Canadian Journal of Earth Sciences* says volcanoes rose with fiery cones about 2.7-2.8 billion years BP.[29] Cratons collided to form the supercontinent Kenorland, which was only the size of Australia and barren with no vegetation. Earth was still a water world with a

superocean called Lerova. The great tides of a closer moon caused immense waves to wash over the land. Vast amounts of open, warm water combined with rapid seasonal temperature changes created powerful monsoons.[30]

Kenorland Monsoon on Barren Earth, imagined from NASA image of tropical Cyclone Veronica, File:Veronica_2019-03-21_0219Z.jpg.

The monsoons produced incredible amounts of rain that quickly eroded the Algoman Mountains from a geological view of time. Sediment flooded the sea with mineral-rich nutrients, and it is hypothesized that green began to appear in the shallow continental margin waters of Lerova. In an article about Earth's oxygen for the *New York Times*, Carl Zimmer says the great marine algae-like bloom of Kenorland is called the Great Oxygen Event, dated to approximately 2.45 billion BP. The Great Oxygen Event is considered essential to the development of life on the planet.[31] In his journal article "Evolution of Minerals" for *Scientific American*, Robert Hazen says it is estimated that more than 2,500 new minerals were created from the approximately 4,400 minerals found on Earth through direct or indirect biochemical reactions.[32]

Evidence of cyanobacteria photosynthesis has been elusive, given the lack of fossils and the age of those previously mentioned. However, Grace Wade, reporting for *New Scientist*, says Emmanuelle Javaux at the University of Liège in Belgium and her colleagues used an electron microscope to analyze fossils from cratons in Australia, Africa, and Greenland. They found evidence of thylakoids, membrane sacs that hold packets of chlorophyll molecules and date the oldest to 1.75 billion BP, closing the gap toward proving a definitive role of

---

[25] Allen P. Nutman et al., "Rapid emergence of life shown by discovery of 3,700-million-year-old microbial structures," *Nature*, 537 (2016): abstract, https://doi.org/10.1038/nature19355.
[26] Vincenzo Rizzo and Nicola Cantasano, "Structural parallels between terrestrial microbialites and Martian sediments: are all cases of 'Pareidolia'?" *International Journal of Astrobiology*, 16, 4 (2016): abstract, https://doi.org/10.1017/S1473550416000355.
[27] Anil Kumar et al., "Evidence for a Neoarchean LIP in the Singhbhum craton, eastern India: implications to Vaalbara supercontinent," *Precambrian Research*, 292 (2017): 17, https://doi.org/10.1016/j.precamres.2017.01.0183.
[28] S. Mahapatro, Naresh Pant, Santanu Bhowmik, A. Tripathy, and Jayanta Nanda, "Archaean granulite facies metamorphism at the Singhbhum Craton-Eastern Ghats Mobile Belt interface: Implication for the Ur supercontinent assembly," *Geological Journal*, 47, 2-3 (2012): abstract, https://doi.org/10.1002/gj.1311.
[29] Charles Gower and Paul Clifford, "The structural geometry and geological history of Archean rocks at Kenora, north-western Ontario-a proposed type area for the Kenoran Orogeny," *Canadian Journal of Earth Sciences*, 18 (2011): abstract, doi.org/10.1139/e81-103.
[30] "What is a Supercontinent And a Superocean?" Worldatlas.com, August 16, 2017, https://www.worldatlas.com/articles/what-is-a-supercontinent-and-a-superocean.html.
[31] Carl Zimmer, "Earth's Oxygen: A Mystery Easy to Take for Granted," *The New York Times*, October 3, 2013, https://www.nytimes.com/2013/10/03/science/earths-oxygen-a-mystery-easy-to-take-for-granted.html.
[32] Robert M. Hazen, "Evolution of Minerals," *Scientific American*, 302, 3 (2010), 63, https://www.scientificamerican.com/article/evolution-of-minerals/.

cyanobacteria in the Great Oxygen Event. She quotes Javaux: "[P]erhaps, they invented thylakoids at this time."[33] In an article for *Live Science*, Mindy Weisberger says research identified a possible reason for the Great Oxygen Event. The early earth spun much faster, and days were just a few hours. Throughout billions of years, the spin slowly decreased. It is speculated that thresholds were reached when more sunlight increased the release of oxygen molecules in the thick cyanobacteria mats.[34] Present-day cyanobacteria blooms in shallow water near the Fiji Islands show a greening similar to that which might have occurred during the Great Oxygen Event.

Filamentous Cyanobacteria Bloom near Fiji, Norman Kuring, 2010, image from NASA's Aqua satellite, File:2010_Filamentous_Cyanobacteria_Bloom_near_Fiji.jpg

More oxygen caused the first extinction. Anaerobic bacteria died off and survive today in the recesses of the earth where no oxygen is present. The cyanobacteria survived through another miraculous event. Daniel Vanchard, in an article on cyanobacteria for the *Encyclopedia of Geology*, says cyanobacteria developed the ability to compensate for the lack of carbon dioxide by using nitrogen to attach the nutrients they needed to grow. The technical term is heterocyst.[35]

In an article for the *Encyclopedia of Astrobiology*, Andrey Bekker wrote that oxygen is thought to have caused a massive glaciation.[36] Oxygen combined with the methane gas of early earth and formed carbon dioxide and water. Carbon dioxide does not retain heat as well as methane.[37] Bekker says the glaciation lasted sixty million years, the longest of Earth's glacial periods. It was named the Huronian Glaciation after discoveries in 2.5-to 2.2-billion-year-old glacial deposits near Lake Huron in Canada between Sault Ste. Marie, Sudbury, and Cobalt.[38] However, the cyanobacteria may be innocent. In an article for *NewScientist,* Stephen Battersby says that the ice may have been caused by something sinister from space, such as an interstellar dust cloud.[39]

The Huronian is thought to be the first snowball Earth period. Glaciation slowed the growth of cyanobacteria, which survived where light penetrated.[40] Brandon Specktor reported in a *Live Science* article that the ice may have retreated in response to a meteor that struck the earth 2.2 billion years ago and left a forty-five-mile-wide crater. All that remains is Barlangi Hill, a small red mount in the Yarrabubba Australian Outback. It hypothesized that if Australia was covered with ice at the time, the amount of water vapor released could have caused the planet to heat up and end the glaciation. Water vapor is a more efficient greenhouse gas than carbon dioxide.[41]

After the glaciers receded, melt and runoff flooded the sea with mineral nutrients. As the invigorated cyanobacteria bloomed, the seas in shallow coastal waters turned green again, and the retreating ice revealed a sandy plain of eroded land. R. W. Ojakangas and fellow researchers write in their article for *Sedimentary Geology* that ocean water flowed into parts of Minnesota, northern Wisconsin, and northern Michigan and

---

[33] Grace Wade, "1.75-billion-year-old fossils help explain how photosynthesis evolved," *New Scientist*, January 3, 2024, https://www.newscientist.com/article/2410391-1-75-billion-year-old-fossils-help-explain-how-photosynthesis-evolved/.
[34] Mindy Weisberger, "Slowdown of Earth's spin caused an oxygen surge," *Live Science*, August 2, 2021, https://www.livescience.com/early-earth-rotation-increase-oxygen.html.
[35] Daniel Vanchard, "Cyanobacteria" in *Encyclopedia of Geology*, 2nd ed. (Amsterdam: Elsevier Ltd, 2021), introduction, https://doi.org/10.1016/B978-0-12-409548-9.11843-3.
[36] Andrey Bekker, "Huronian Glaciation," in *Encyclopedia of Astrobiology* (Berlin, Heidelberg: Springer, 2011), 772, https://doi.org/10.1007/978-3-642-11274-4_742.
[37] Princeton University, "A more potent greenhouse gas than carbon dioxide, methane emissions will leap as Earth warms," *ScienceDaily,* March 27, 2014, https://www.sciencedaily.com/releases/2014/03/140327111724.htm.
[38] Andrey Bekker, "Huronian Glaciation," 768-769
[39] Stephen Battersby, "Earth's wild ride: Our voyage through the Milky Way," *NewScientist,* November 30, 2011, https://www.newscientist.com/article/mg21228411-500-earths-wild-ride-our-voyage-through-the-milky-way/.
[40] University of California-Davis, "Oxygen oasis in Antarctic lake reflects Earth in distant past," *ScienceDaily,* September 1, 2015, https://www.sciencedaily.com/releases/2015/09/150901140759.htm.
[41] Brandon Specktor, "Earth's oldest known meteor crash site found in Australian Outback," *Live Science,* January 22, 2020, https://www.livescience.com/worlds-oldest-meteor-crater-yarrabubba.html.

formed the epi-continental Animikie Sea.[42] Cyanobacteria flourished in the shallow water. Oxygen interacted with dissolved iron from eroded volcanic rock and precipitated iron oxide that fell to the seafloor. Layer after layer of sediment and iron created banded iron deposits in an arc from Minnesota through the Lake Superior region into Michigan. The Minnesota range is called the Mesabi. The sixty-mile range in Wisconsin and Michigan is called the Gogebic. This process occurred worldwide in shallow seas.[43] Iron in seawater is a critical element for almost every living cell.

As cyanobacteria flourished in the Animikie Sea, the shape of North America began to form with a massive super mountain event called the Trans-Hudson. Marc R. St-Onge and colleagues, in an article for *Techtonics*, date the event at 1.83 BP. The range spanned 4,600 kilometers from the Dakotas to Hudson Bay. The Black Hills of South Dakota are a remnant. They say the Trans-Hudson orogeny is comparable to the Himalayas.[44]

Toward the end of the mountain building, the earth plunged into darkness. In an article for *Live Science*, Becky Oskin wrote that a comet struck in Canada, the Sudbury impact, the second largest crater to be discovered. Dated to 1.85 billion years BP, rock fragments were ejected as far as Minnesota, with an impact crater estimated to be 160 miles wide.[45] Writing for *Wired Science*, Sid Perkins says the impact spread debris globally. The event ended the iron band accumulation in the Animikie Sea.[46]

The sky barely cleared when it darkened again with the smoke of volcanoes. Klaus Schultz and William Cannon say in their *Precambrian Research* article that the Penokean orogeny is dated to 1.88 billion years BP, when land terranes collided with the Superior Craton and covered the Ocooch Mountain area, Minnesota, upper Wisconsin, the Lake Superior region, and upper Michigan.[47] Some peaks may have been as high as the Alps. Only a fragment remains. Mt Whittlesey rises to 1,872 feet.[48]

Mt. Whittlesey. Photo courtesy of Joel Austin, author of *Discovering the Penokees*.

Martin compares the Penokean Range to the Blue Ridge Mountains in the Appalachians, although not as high. The Penokean remnant roots are a mere twenty-five to thirty miles long compared to the massive range that covered multiple states. Martin writes that the Penokean Range is unusual for numerous ancient water gaps where rivers flowed between peaks. He says nine water gaps exist between the Montreal River and Bad River. The Blue Ridge Mountains have half this number in a comparable distance, and there are thirty-mile areas "with no gaps whatsoever."[49]

After millions of years, incredible amounts of quartz sand eroded from the Penokean Mountains and accumulated as ancient seas ebbed and flowed in the Ocooch Mountain area. Richard A. Davis Jr. and researchers in *Geology of the Baraboo* date a massive surge of lava that created the Baraboo Hills at approximately 1.6 billion years BP.[50] Wave ripples in the seashore sand frozen in quantize can be seen at Devil's Lake State Park. Martin says the Paleozoic Plateau sediment once covered the ancient hills.[51]

At 1.1 billion years BP, the deeply buried Great Lake Tectonic Zone, GLTZ, an ancient crustal boundary, erupted with massive force. A great rift in the crust ripped open in the Lake Superior region. Rifts are part of

---

[42] R. W. Ojakangas et al., "Paleoproterozoic basin development and sedimentation in the Lake Superior region, North America," *Sedimentary Geology*, 141–142 (2001): abstract, https://doi.org/10.1016/S0037-0738(01)00081-1.

[43] Universitaet Tübingen, "Iron in primeval seas rusted by bacteria," *ScienceDaily*, April 23, 2013, https://www.sciencedaily.com/releases/2013/04/130423110750.htm.

[44] Marc R. St-Onge, et al., "Hudson Orogen of North America and Himalaya-Karakoram-Tibetan Orogen of Asia: Structural and thermal characteristics of the lower and upper plates," *Techtonics*, an AGU Journal, 25, 4 (July 18, 2006): [5], https://doi.org/10.1029/2005TC001907.

[45] Becky Oskin, "Crash! 10 Biggest Impact Craters on Earth," *Live Science*, April 28, 2014, https://www.livescience.com/45126-biggest-impact-crater-earth-countdown.html.

[46] Sid Perkins, "Giant Asteroid Impact Could Have Stirred Entire Ocean," *Wired Science*, November 10, 2009, https://www.wired.com/2009/11/giant-asteroid-impact-could-have-stirred-entire-ocean/.

[47] Klaus J. Schultz and William F. Cannon, "The Penokean orogeny in the Lake Superior region," *Precambrian Research*, 157, 1–4 (2007): abstract, introduction, https://doi.org/10.1016/j.precamres.2007.02.022.

[48] "Mt. Whittlesey," Peakery, accessed July 15, 2019, https://peakery.com/mount-whittlesey-wisconsin/.

[49] Martin, *The Physical Geography of Wisconsin*, 377, 379

[50] Richard A. Davis Jr., Robert H. Dott, Jr., and Ian W. D. Dalziel, "Proterozoic geology of the Baraboo Interval," in *Geology of the Baraboo, Wisconsin, Area,* 43 (Boulder: Geological Society of America, January 1, 2016), 11, https://doi.org/10.1130/2016.0043(05).

[51] Martin, *The Physical Geography of Wisconsin*, 38

the breakup cycle of plate tectonics. Usually, the ocean flows into the rift area as the crustal pieces separate. Jessica Marshall reports for *Nature News* that the mid-continent rift behaved differently: "It opened a 3,000-kilometer crack in North America and created a basin as big, perhaps, as the Red Sea-then the system shut down." Marshall quotes G. Randy Keller, a geophysicist who describes the event's magnitude: "How that feature could just totally reorganize the crust of the Earth in the Lake Superior region and not manage to break the continent apart is fairly amazing…It's a spectacular failure."[52]

Marshall says geophysicists determined the rift has the shape of a horseshoe with two ends that point south from Lake Superior. It is believed the lava, called basalt, pushed upward from the force of a hot mantle plume and rose to the surface with such power that seismic studies in the 1980s measured rock almost nineteen miles deep: "All told, the rift produced between 1 million and 2 million cubic kilometers of basalt, making it one of the world's largest deposits of that rock." Marshall also quotes Nicholas Swanson-Hysell, a geologist at the University of California, Berkeley, who describes the pristine nature of the lava flows protected by the stable Superior Province: "It's gorgeous…How well these flows are preserved is pretty amazing. You could go to a lava field in Hawaii that erupted in 1950, and the surface would look similar to this 1.1-billion-year-old surface."[53]

The immense amount of lava in the zone interacted with sediments and created copper deposits near the southern shores of Lake Superior. Isaac Orr, writing for American Experiment, says Minnesota has the largest untapped copper-nickel deposit in the world.[54] Recent glaciation exposed copper deposits and scattered pieces on the eastern side of Wisconsin as erratic boulders. This enabled the Native American Old Copper Complex to develop throughout Wisconsin and the Ocooch Mountains. Complex refers to multiple traditions that used copper to form tools and ornaments. Additionally, minerals filled volcanic gas bubble cavities, and glaciers scrubbed them off as agates found on the shores of Lake Superior and Lake Michigan.

A 1985 study by Schmus and Hinze in the *Annual Review of Earth and Planetary Science* suggests a crustal collision may have sutured the deep rift.[55] The suture source may have been the massive Grenville Orogeny, which Richard Tollo and fellow researchers discussed in a Geological Society of America publication. This orogeny raised mountains approximately 1.3-1 billion years BP from Mexico to Labrador, including the Blue Ridge, Appalachians, and Adirondacks.[56] The Grenville Orogeny was part of a worldwide mountain building associated with the supercontinent Rodinia. This kept the Ocooch Mountain area connected to North America.

With all the heat and fire of the Grenville Orogeny, it is hard to believe glaciers followed and created another snowball earth! Two periods of glaciation occurred at 720 and 640 million years BP, appropriately called the Cryogenic Age.[57] In an article featured on NASA's news site, Michael Schirber says the primary evidence for these two snowball earth periods is glacier debris near the equator. Each episode lasted approximately ten million years, and there is debate on whether the freeze was total or "slushy."[58] Penn State News reported in a *ScienceDaily* article that lichens may have played a role. They are a composite of cyanobacteria or algae and fungus filaments that emerged on land. Lichens produce acids strong enough to dissolve limestone rocks and create calcium carbonate. Washed into the ocean, carbon atoms cannot form the carbon dioxide greenhouse gas in the atmosphere. Trapped carbon may have enabled the glaciers to form.[59]

J. M. Archibald's article in the *Encyclopedia of Microbiology* says algae formed when cyanobacteria entered cells that had DNA.[60] The cyanobacteria provided the host cell with critical nutrients (organic compounds), which eventually enabled the host cell to develop the ability of photosynthesis in return for a home. The appearance of plants, together with cyanobacteria, may have produced enough oxygen to spark the explosion of life known as the Cambrian Period. All animal forms of life we have today can be traced back to this

---

[52] Jessica Marshall, "Geology: North America's broken heart" *Nature News Feature*, December 4, 2013. https://www.nature.com/news/geology-north-america-s-broken-heart-1.14281.
[53] Marshall, "Geology: North America's broken heart"
[54] Isaac Orr, "Minnesota's Copper and Nickel Deposits are World Class," Americanexperiment, March 12, 2018, https://www.americanexperiment.org/2018/03/minnesotas-copper-nickel-deposits-world-class/.
[55] W. R. Schmus and W. J. Hinze, "The Midcontinent Rift System," *Annual Review of Earth and Planetary Science* (May 13, 1985), 379, doi: 10.1146/annurev.ea.13.050185.00202.
[56] Richard Tollo et al., "Proterozoic tectonic evolution of the Grenville orogen" in *North America: An Introduction, Memoir of the Geological Society of America,* 197 (Boulder: Geological Society of America, September 1, 2004), 1-14.
[57] "Chart/Time Scale," International Commission on Stratigraphy, accessed July 15, 2019, https://web.archive.org/web/20170113013553/http://www.stratigraphy.org/index.php/ics-chart-timescale.
[58] Michael Schirber, "Snowball Earth' Might Have Been Slushy," NASA, Research Features, accessed January 14th, 2023, https://www.giss.nasa.gov/research/features/201508_slushball/.
[59] Penn State, "First Land Plants And Fungi Changed Earth's Climate, Paving The Way For Explosive Evolution Of Land Animals, New Gene Study Suggests," *ScienceDaily*, August 10, 2001, www.sciencedaily.com/releases/2001/08/010810070021.htm.
[60] J. M. Archibald, "Secondary Endosymbiosis" in *Encyclopedia of Microbiology*, 3rd ed. (Amsterdam: Elsevier, 2009), abstract, accessed July 15, 2019, https://doi.org/10.1016/B978-012373944-5.00360-6.

miraculous era. Gene mutation occurs at regular intervals that can be calculated like a molecular clock to reveal evolutionary history. Blair Hedges, an evolutionary biologist, is quoted in the *ScienceDaily* article on Penn State News: "The plants conceivably boosted oxygen levels in the atmosphere high enough for animals to develop skeletons, grow larger, and diversify."[61] The world at this time is described by Douglas Fox for *Nature*:

The stone formations [stromatolites] are indeed monuments of a faded empire, but not from anything hewn by human hands. They are pinnacle reefs, built by cyanobacteria on the shallow sea floor 543 million years ago, during a time known as the Ediacaran period. The ancient world occupied by these reefs was truly alien. The oceans held so little oxygen that modern fish would quickly flounder and die there. A gooey mat of microbes covered the seafloor at the time, and on that blanket lived a variety of enigmatic animals whose bodies resembled thin, quilted pillows. Most were stationary, but a few meandered blindly over the slime, grazing on the microbes. Animal life at this point was simple, and there were no predators. But an evolutionary storm would soon upend this quiet world.[62]

Enough oxygen produced by cyanobacteria and their plant allies is believed to have breached an ecological threshold, and predators evolved. Fox says: "The rise of carnivory would have set off an evolutionary arms race that led to the burst of complex body types and behaviors that fill the oceans today."[63]

The cyanobacteria's two-billion-year rule began to end as life filled the sea. Evidence is the lack of fossils and the appearance of thrombolites, a mushy version of stromatolites. Rachel Nuwer, in an article for *Smithsonian Magazine Smart News*, writes that genetic sequencing of modern thrombolites showed significant foraminifera, a simple single-cell ocean microbe, and simulations of billion-year-old conditions revealed the presence of foraminifera degraded the fine sedimentary layers of stromatolites.[64] Pristine versions survived to leave us precious unadulterated fossils in the Ocooch Mountain area thanks to the salty shallow epi-continental sea that Sloss could have "walked through." Salt hindered microbe development and organisms that grazed on the bacterial mats.

While life flourished in the oceans, Dr. Dott and Attig wrote: "Strange as it may seem, Cambrian lands may have resembled the surface of Mars more than any earthly habitat."[65] As the sea waxed and waned, the porous soluble limestone of the lower Ordovician Prairie du Chien dolomite accumulated. The topography is karst, from the German word kräs, which refers to a "limestone plateau northeast of the Istrian Peninsula in western Slovenia extending into eastern Italy," where karst was first studied.[66] Karst topography in the Ocooch Mountains is associated with caves and cave systems. Signs of a cave include sinkholes where a cave roof collapsed and disappearing streams where no outlet can be found. The stream may flow into a sinkhole, through fractures to an aquifer, or become an underground stream, which might emerge as a spring or cold stream, an ideal habitat for trout. A modern cave formed south of the Ocooch Mountain area near Blue Mound, the highest point in southern Wisconsin at 1,719 feet.[67] The National Park Service declared Cave of the Mounds a National Landmark.[68] Stromatolite silicified fossil stones that weather up into fields on the top of hills, bluffs,

and ridges may also be a sign of caves. However, as Dr. Dott and his colleagues explained, these colorful and strangely shaped rocks are only present where water was particularly effective in creating cavities and caves and mineralizing the stromatolite fossils. If you find an interesting Ocooch Mountain Rock, you may experience a sense of awe and wonder now that you know cyanobacteria has journeyed through time, fire, and ice. A great-great-grandson or granddaughter may find an Ocooch Mountain Rock on Mars!

NASA, *Curiosity's track on Martian Dune,* 2014, Wikimedia Commons.

---

[61] Penn State, "First Land Plants And Fungi Changed Earth's Climate"
[62] Douglas Fox, "What Sparked the Cambrian explosion?" *Nature News Feature*, February 16, 2016, https://www.nature.com/news/what-sparked-the-cambrian-explosion-1.19379.
[63] Fox, "What Sparked the Cambrian explosion?"
[64] Rachel Nuwer, "What Happened to the Stromatolites, the Most Ancient Visible Lifeforms on Earth?" *Smithsonian Magazine Smart News,* May 30, 2013, https://www.smithsonianmag.com/smart-news/what-happened-to-the-stromatolites-the-most-ancient-visible-lifeforms-on-earth-84714880/.
[65] Dott, Jr. and Attig, *Roadside Geology in Wisconsin,* 10
[66] "Karst," Merriam Webster, July 20, 2020, https://www.merriam-webster.com/dictionary/karst.
[67] "Wisconsin High Points," State Cartographer's Office, University of Wisconsin-Madison, updated: August 14, 2017, https://www.sco.wisc.edu/wisconsin/high-points/.
[68] "Wisconsin," National Park Service, accessed July 15, 2019, https://www.nps.gov/subjects/nnlandmarks/state.htm?State=WI.

A Gallery of Strange, Beautiful, & Scary Rocks!

Many pictures in the gallery have exciting shapes that resemble things. Some have faces or eyes. Once seen, they may never be found again, so I try to capture an intriguing image with whatever cellphone or camera is available. Some rocks are fragile, and all are rare. Those that hold your attention are collector treasures. The ability to see familiar shapes in nature, like clouds or rocks, is called pareidolia, the tendency of people to seek patterns. The texture varies from pumice to glass. The photos are natural art.

**Bowl-like fragments hold water for a day or two.
A bird figurine perches on a stone nest and a horseshoe-like feature.**

The photo shows how Stromatolite impressions may have formed. *Stromatolites in the Pika Formation (Cambrian) near Helen Lake, Banff National Park, Canada.* Photograph taken by Mark A. Wilson (Department of Geology, The College of Wooster), August 8, 2009. Released to public domain, File: CambrianStromatolites.jpg.

Cavities in stromatolite rock might be impressions.

Animal-shape, Figurine, Bird-shape, tail up, Prairie Dog Pups, Stone Creature, Oracle Rock, and Rock Troll stands on a stone perch.

Owl Family in their stone tree!

Stromatolite stem with flat top points to ghost rock.

Can you see the Stone Faces?

Dinosaur Skull Stone Face below

Ghost Rocks in the Snow

Rock with hole, mystical rock, botryoidal, and Christmas tree rock have flared tops.

The Old Stromatolite
A Conophyton

Modern-day stromatolites in Australia
Salty water deters predators. Domes may collapse, creating holes.

Ruth Ellison, *Stromatolites at Lake Thetis, Western Australia,* March 30, 2006, CC 2.0, https://www.flickr.com/photos/laruth/153584043/.

Stromatolite Rocks, the hole is the remaining feature.

Massive Stone Eye has faint sedimentary streaks.

White quartz with empty cavities ringed with purple and red ribbon quartz.

Quartz slabs give a glass-like feel.
Round balls of quartz from a stromatolite cavity,
and black spots cover white-brown and greenish stromatolite rock.

**Stromatolite Fossil Hunt.
A good find for the day!**

Photo by Bryon Bootman, 2020, *Jurassic Park?*

Stone Garden Creature

Scary Rock Fish

Dragon with white frozen fire in stone

Odin Owl

Large & Small
Rock Spiders

Rock Creature

Ancient Head Rock
is blurry but remarkable.
Can you see a face in the right eye?
More follow; all are surreal.

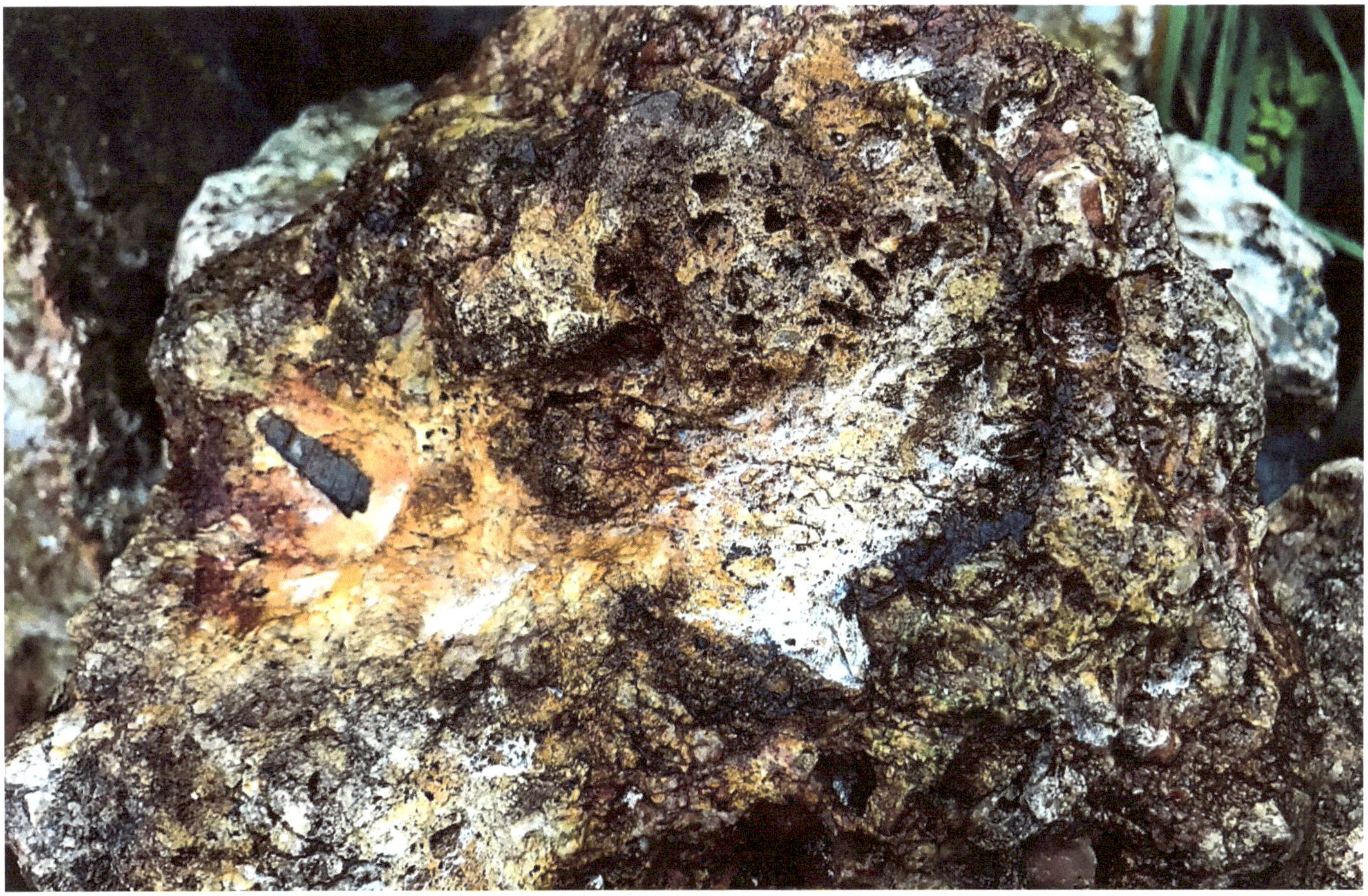

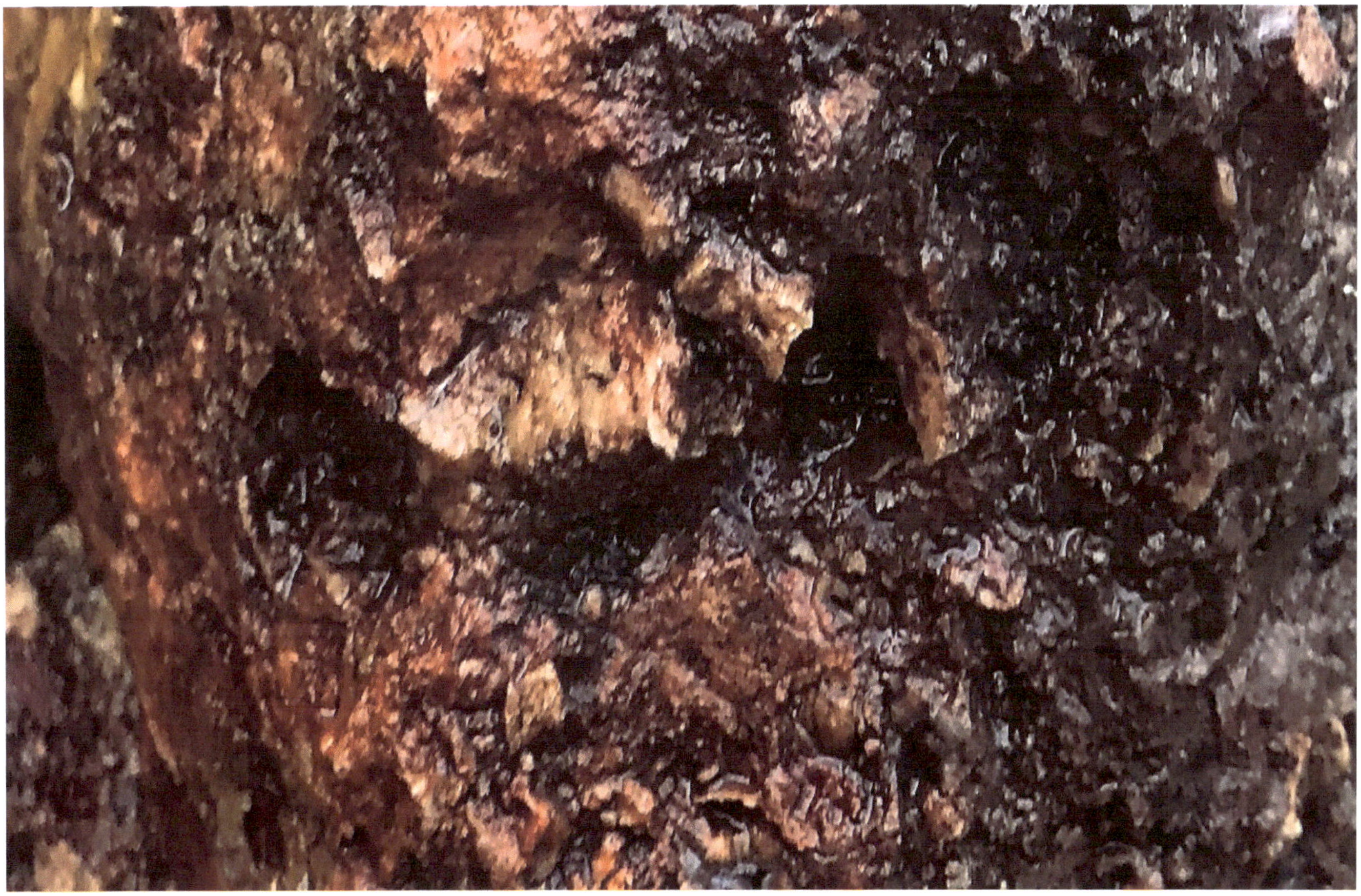

Old Woman and a Baby

Woman with long hair

Stone Horse with many horns!

Dragon Rock
The eye is a slit in grayish quartz.
A Rock Face watches!

April's Owl & Surreal images on Magnified Quartz

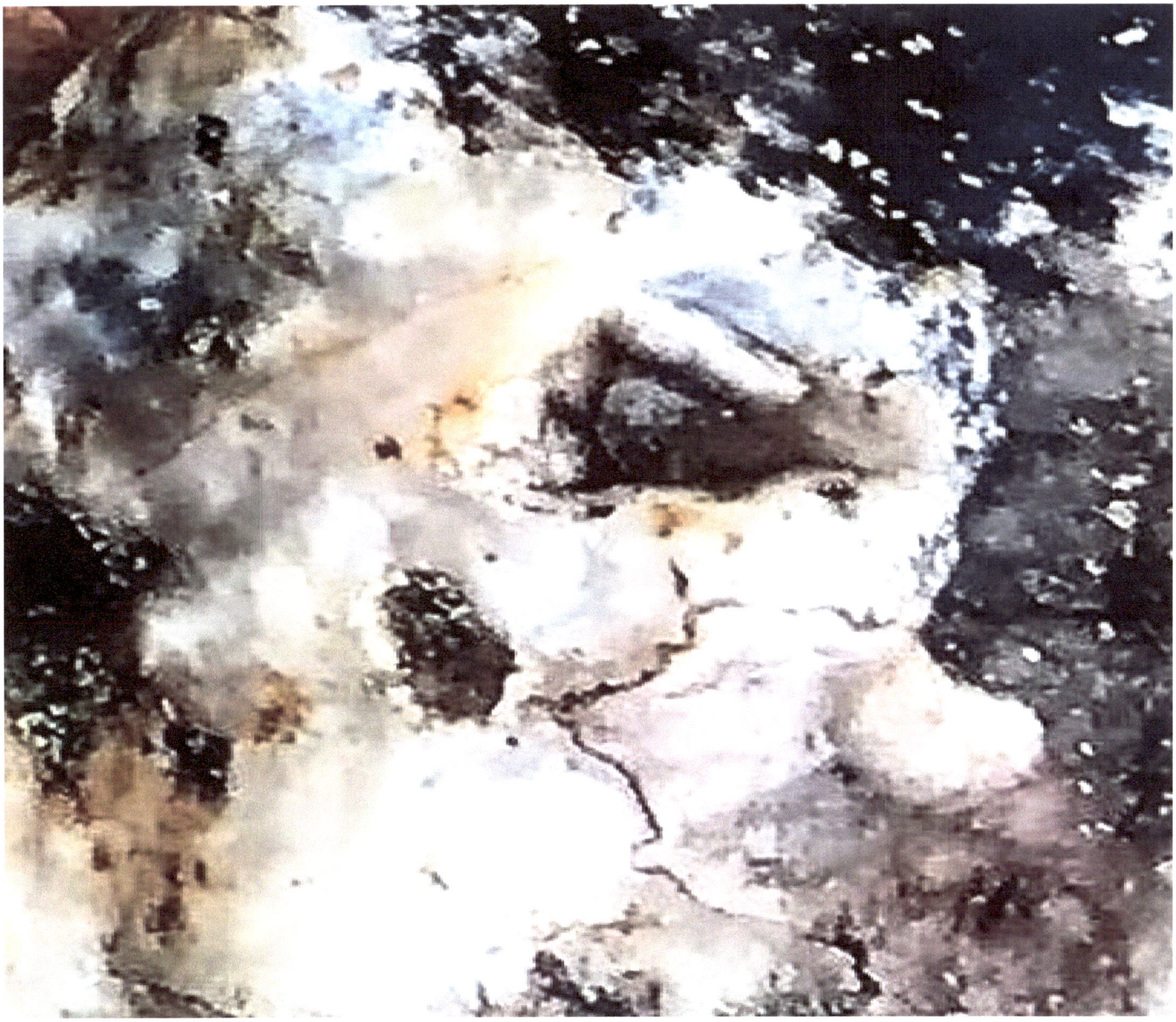

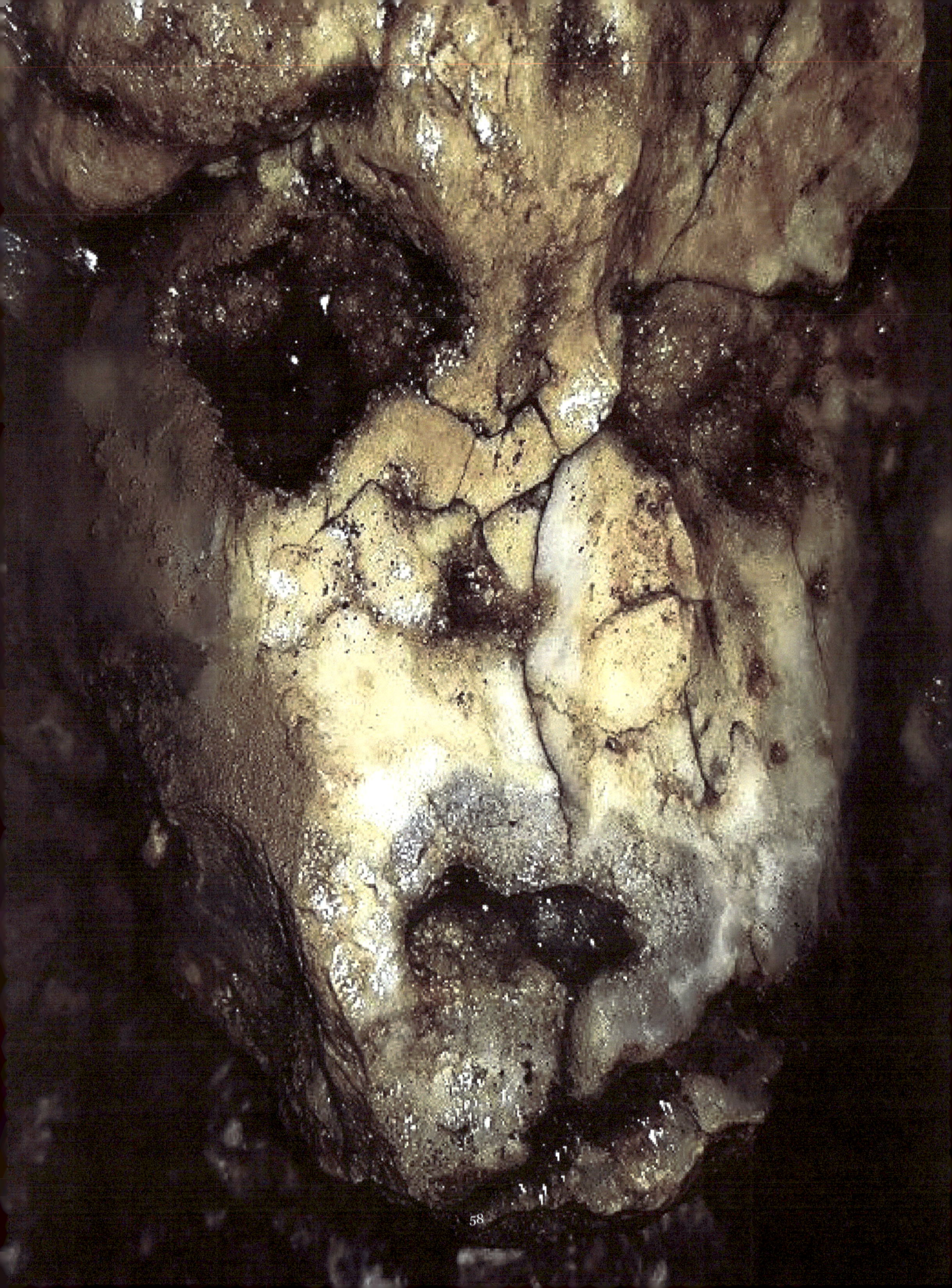

Eric Jonathan Zingler

Collage of Ancient Rock Faces

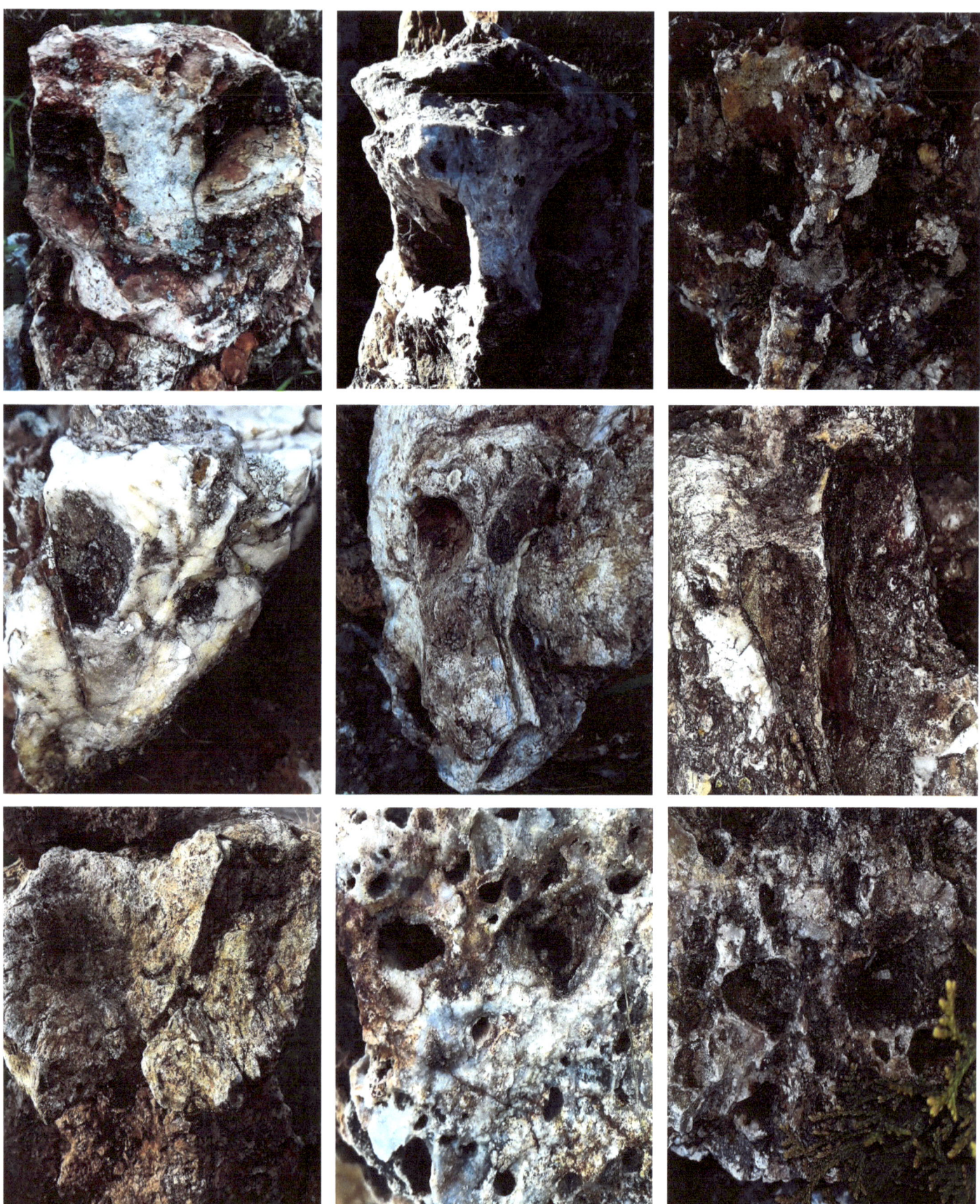

Strange faces!

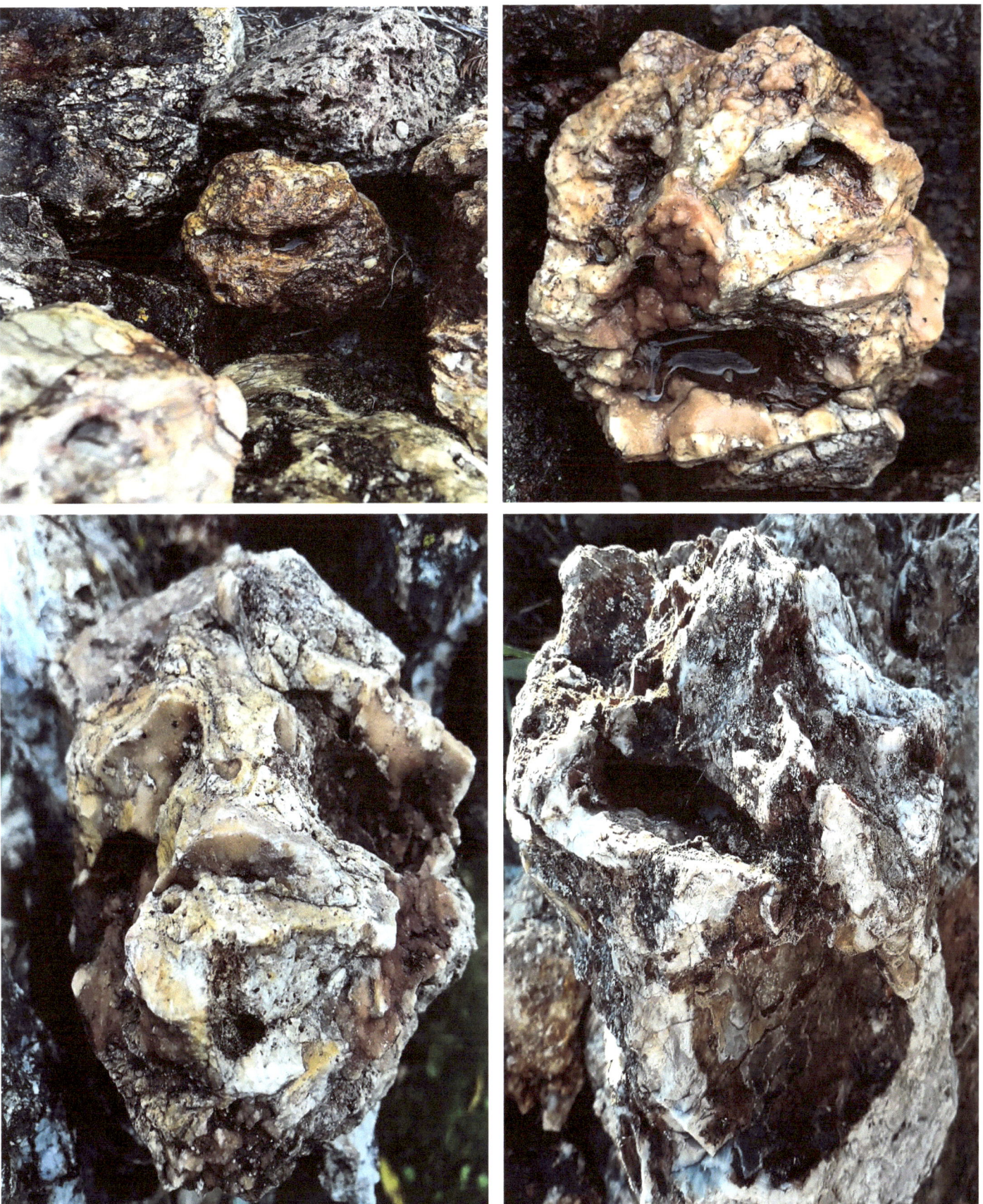

Rock Voices

Faces or Eyes in Stone Blocks of Stromatolites

Eyes are characteristic of Ocooch Mountain Rocks

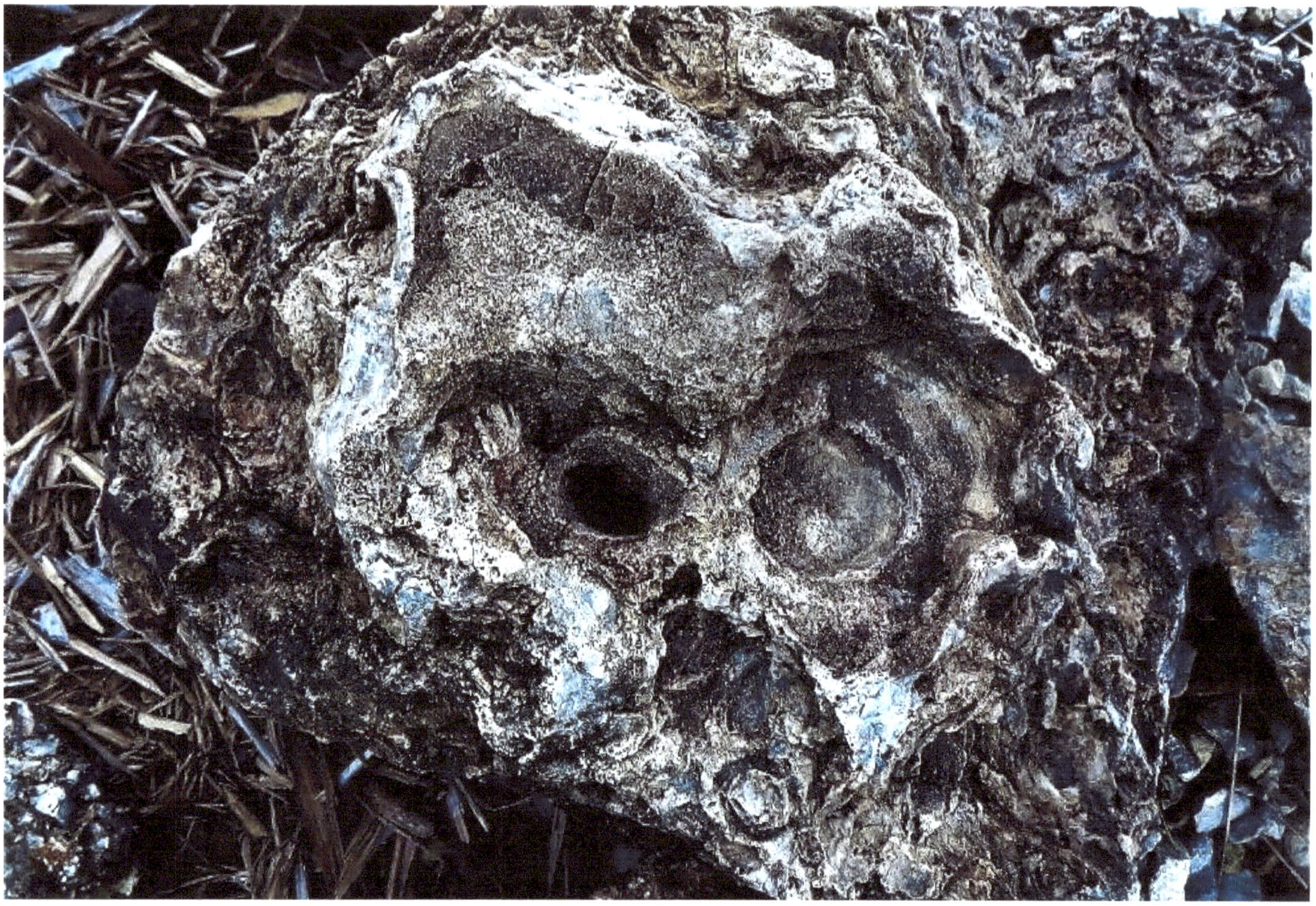

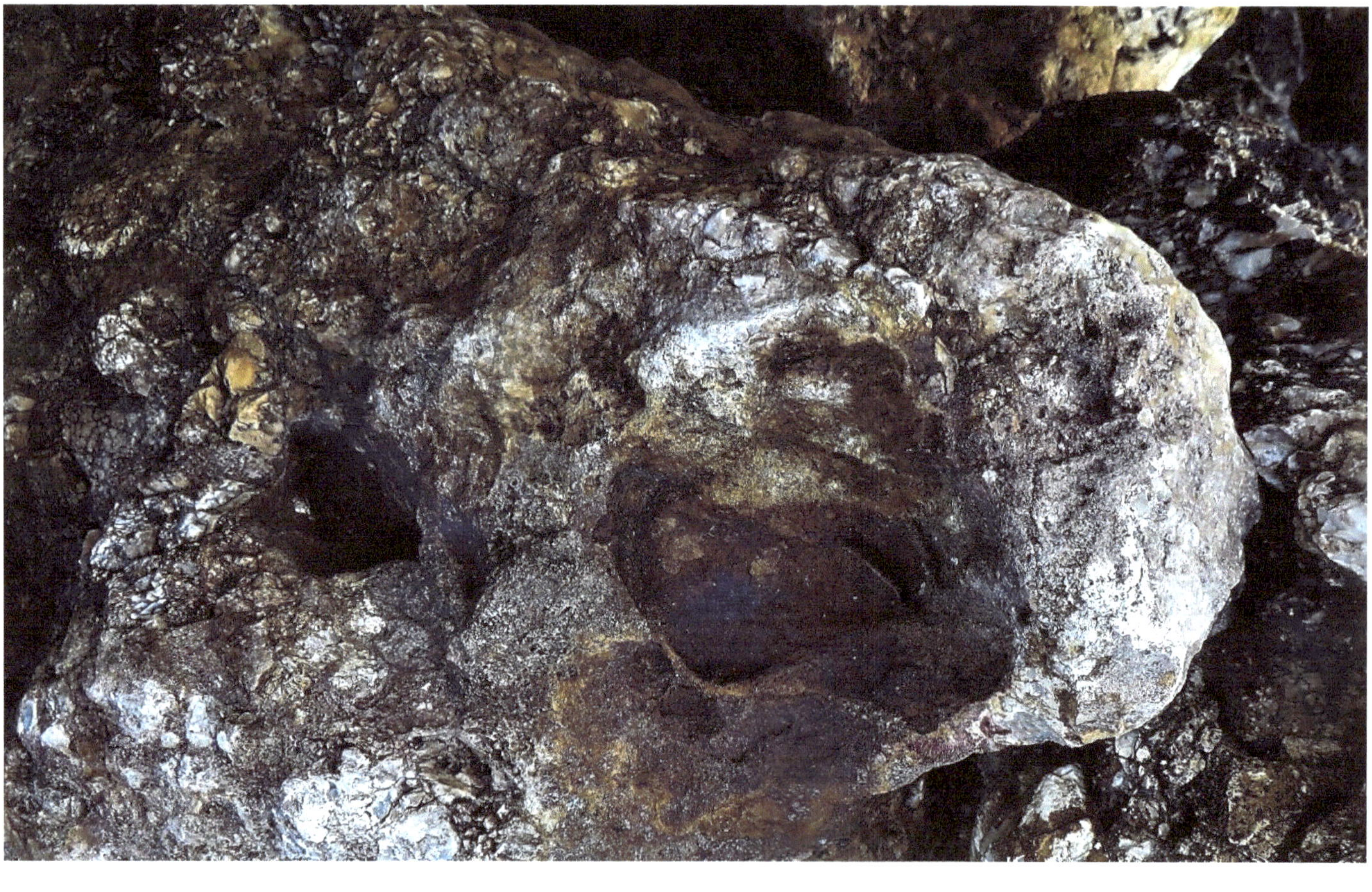

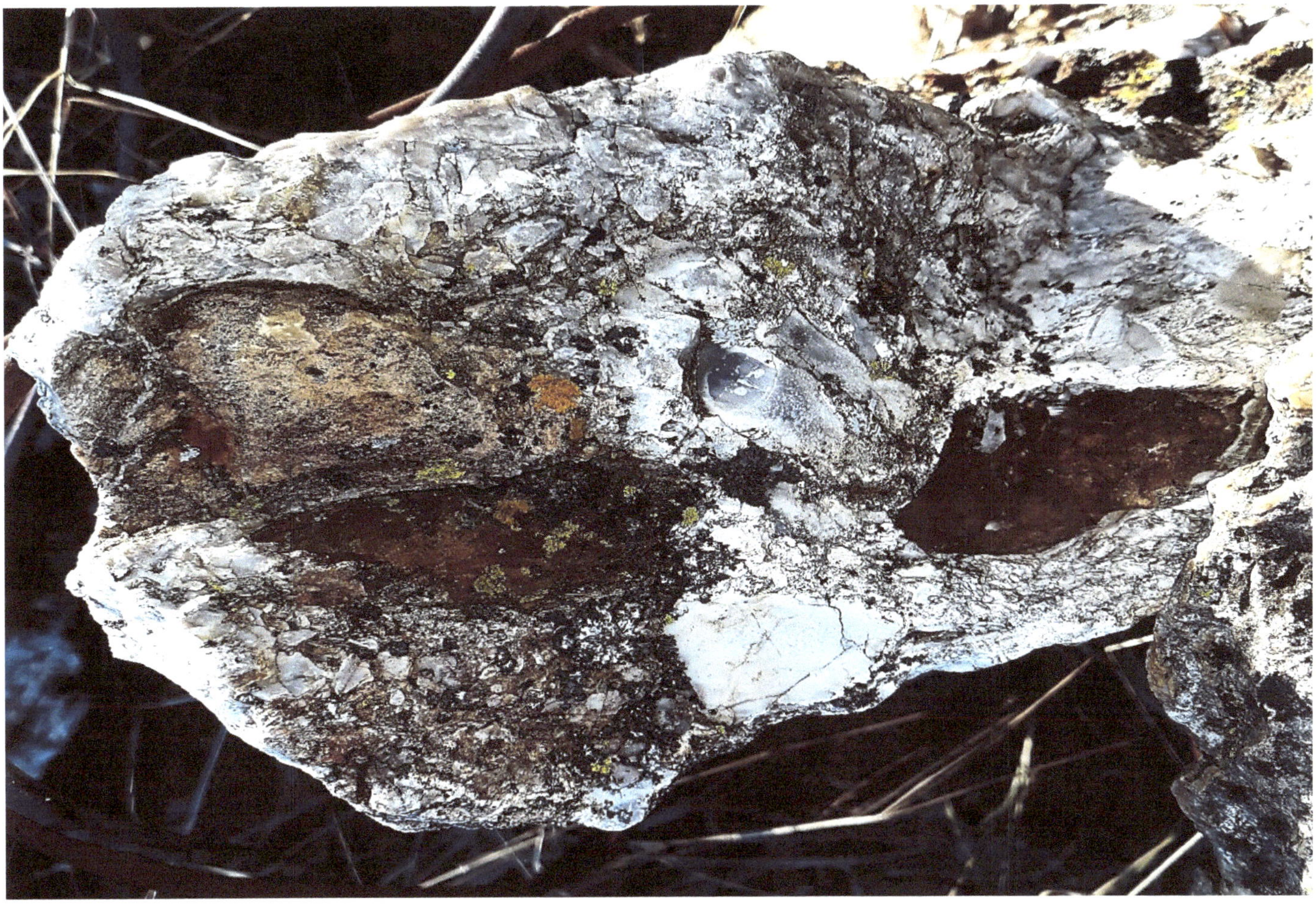

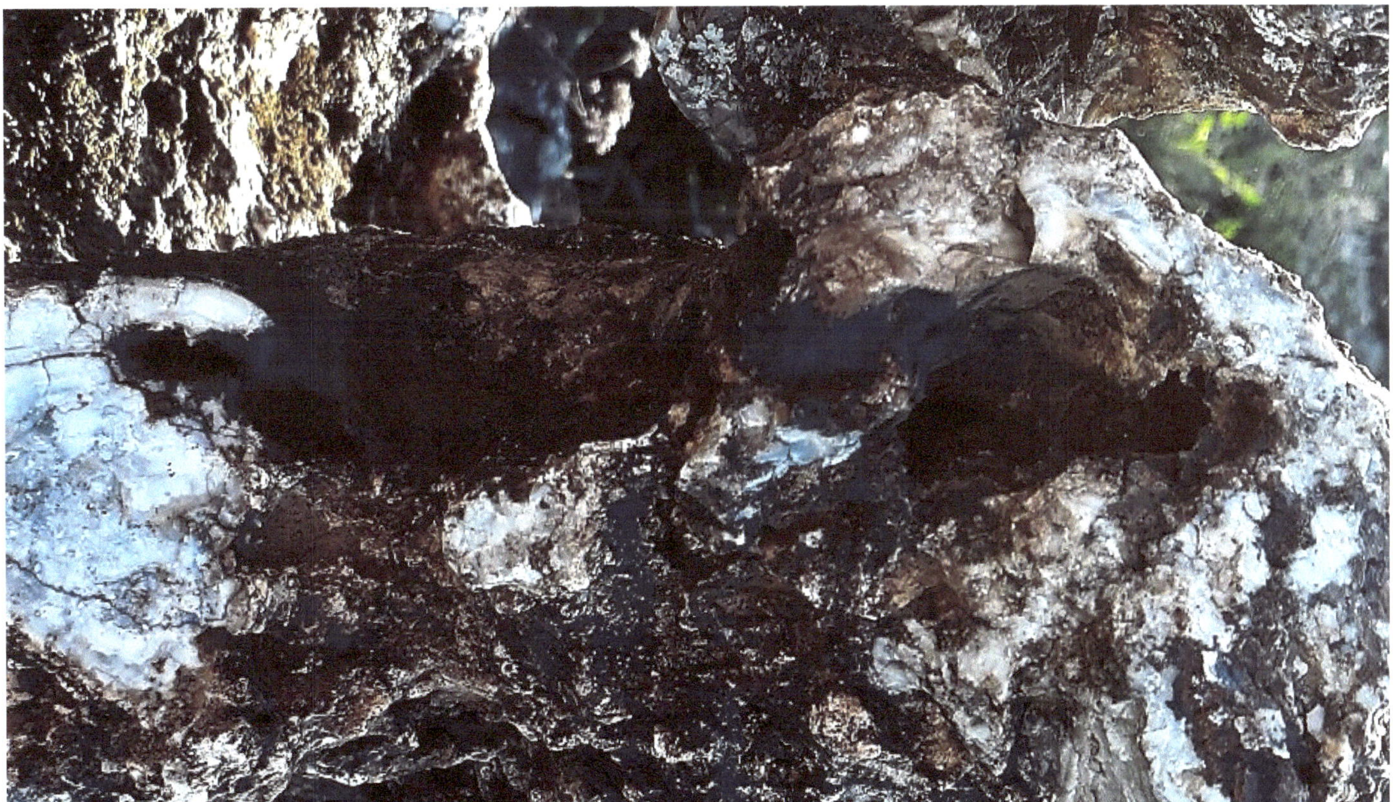

Eye of the World

Rocks that Stare

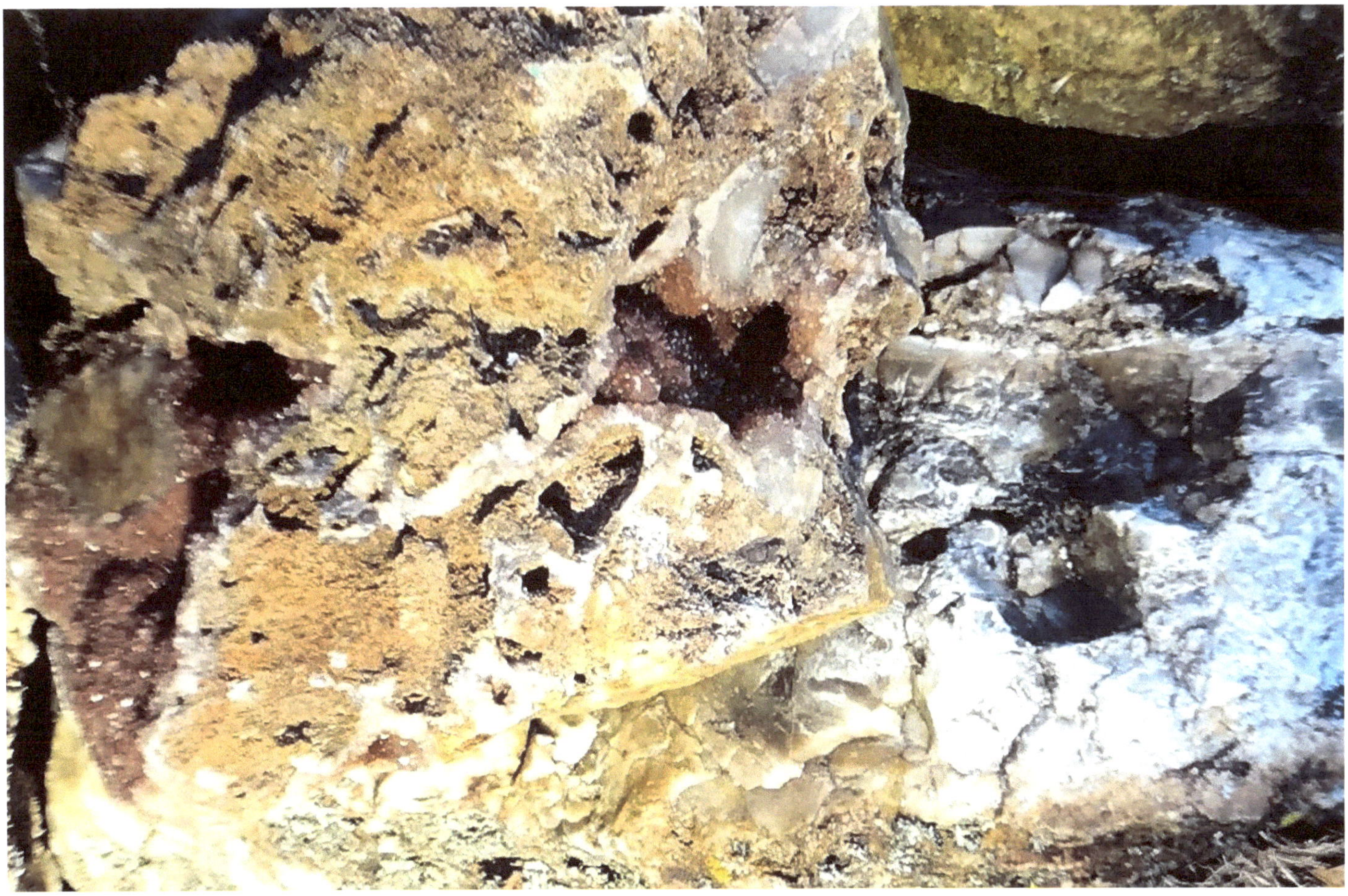

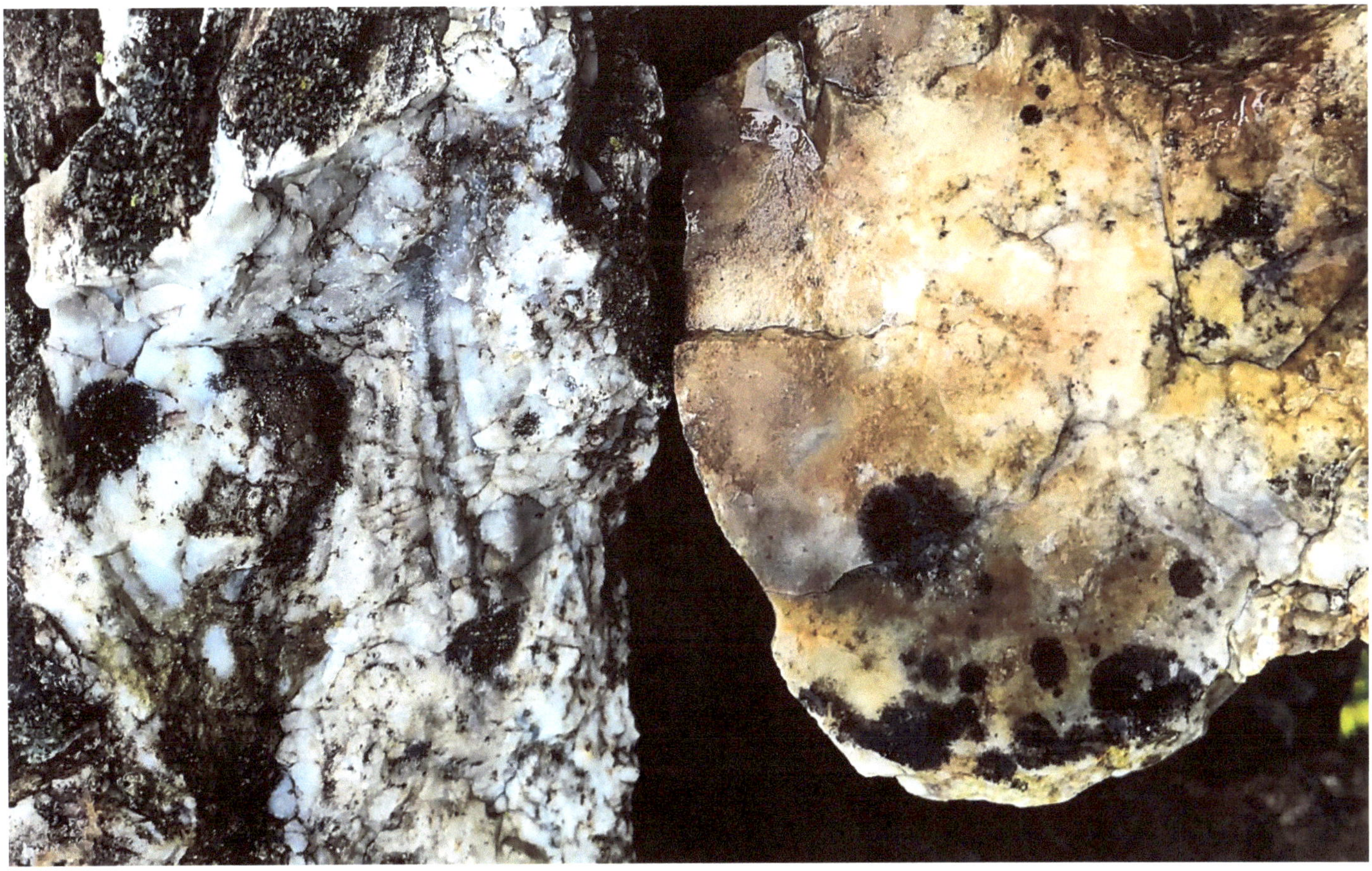

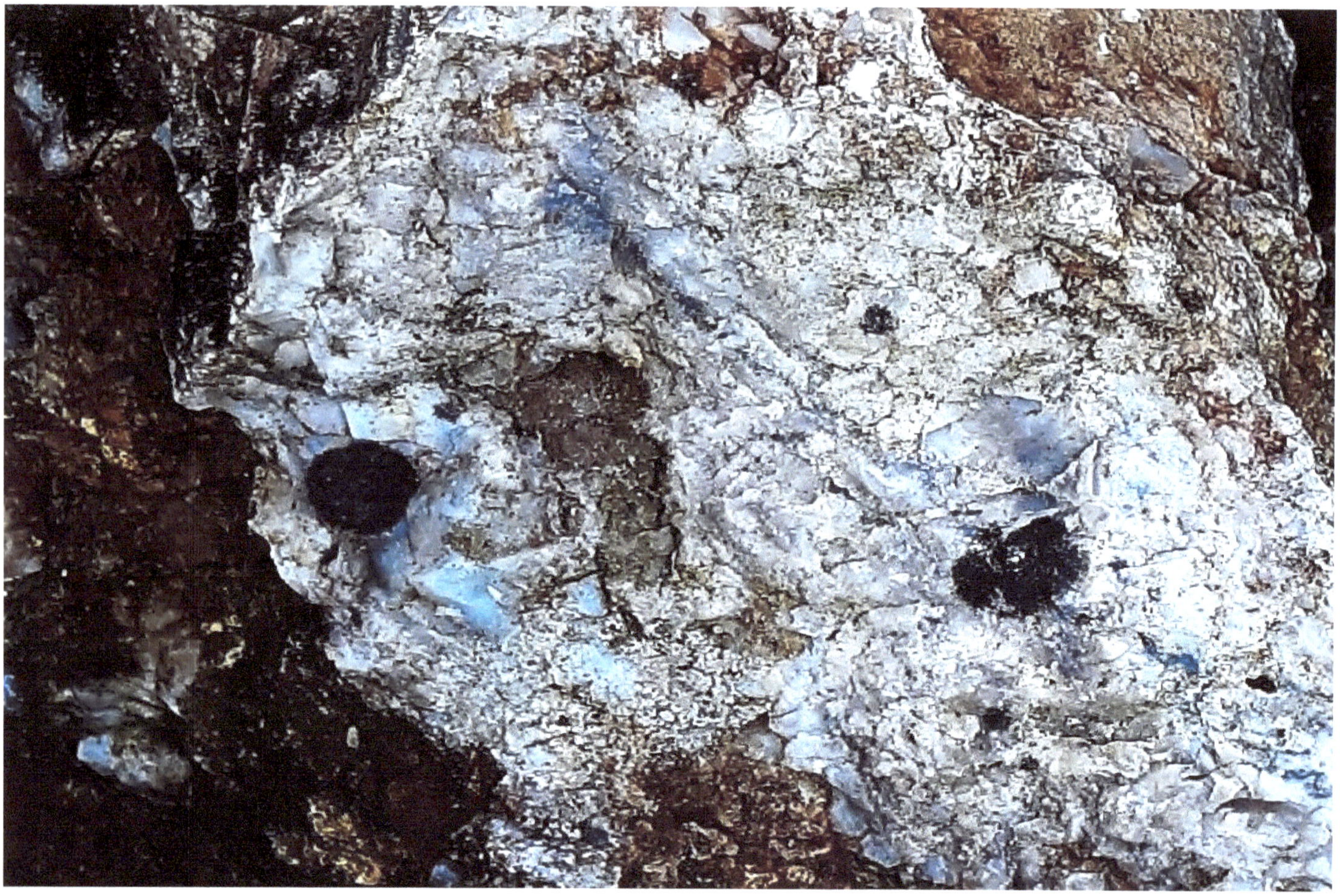

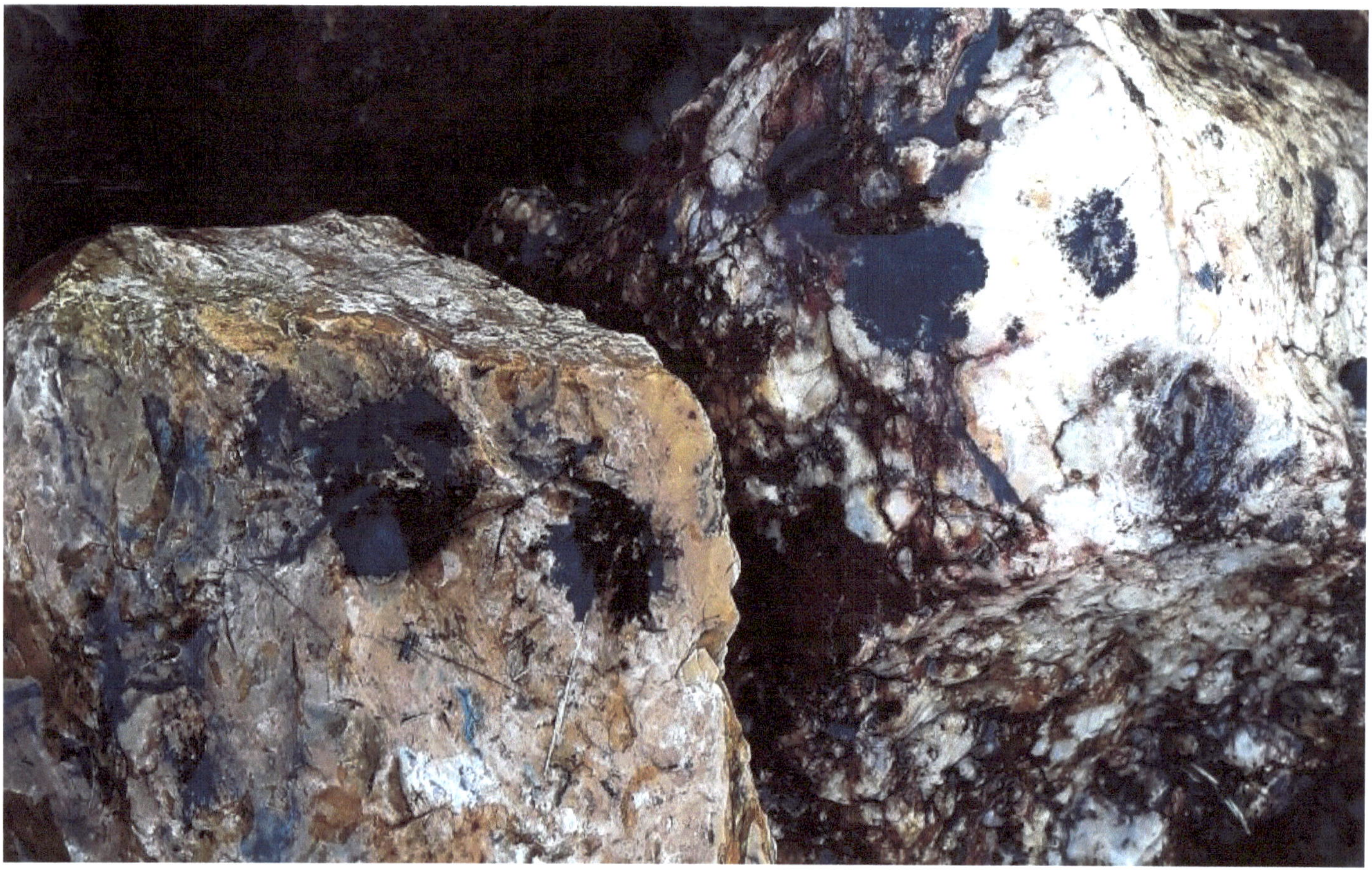

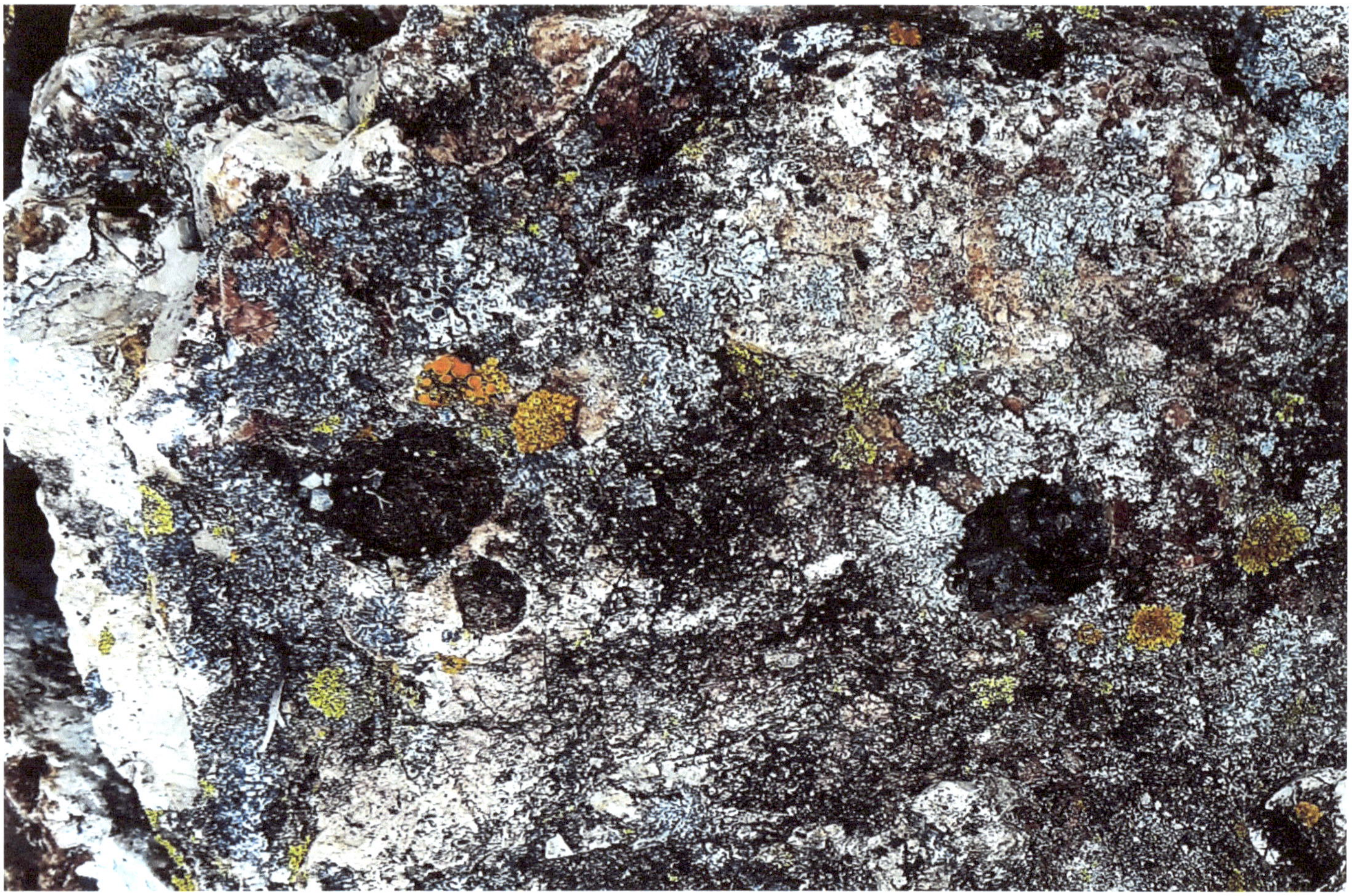

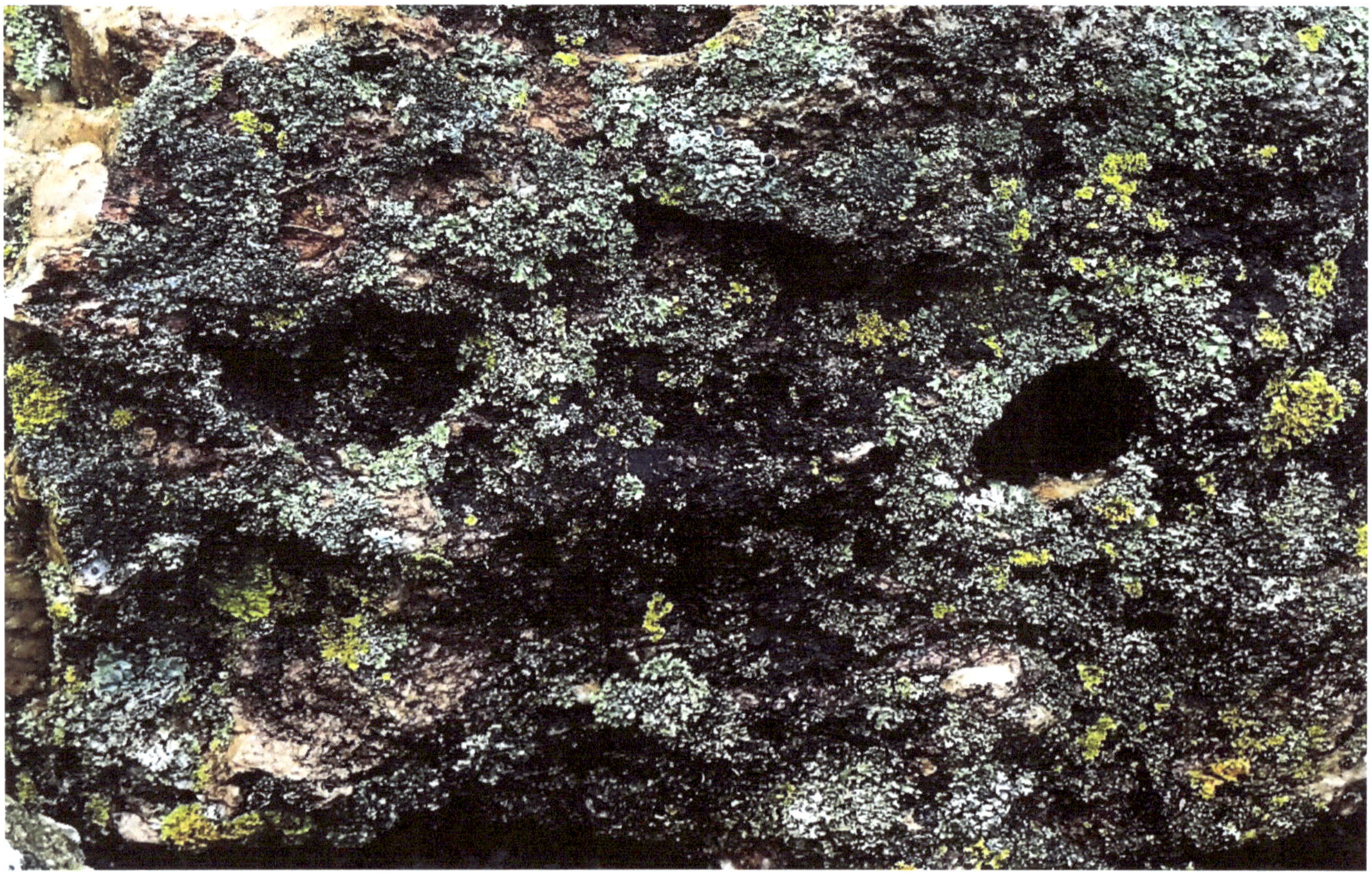

Skull Rocks!
Note the teeth!

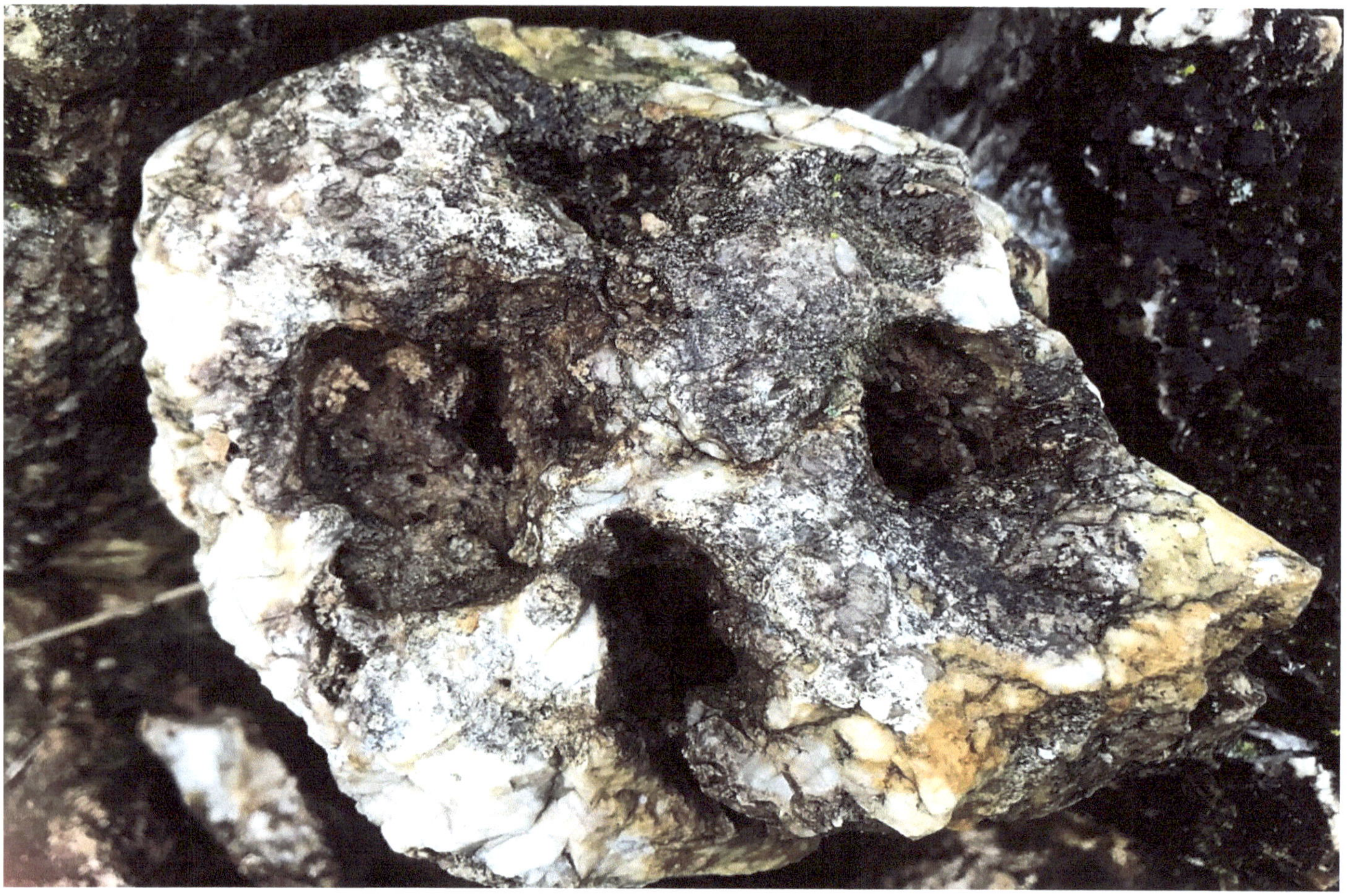

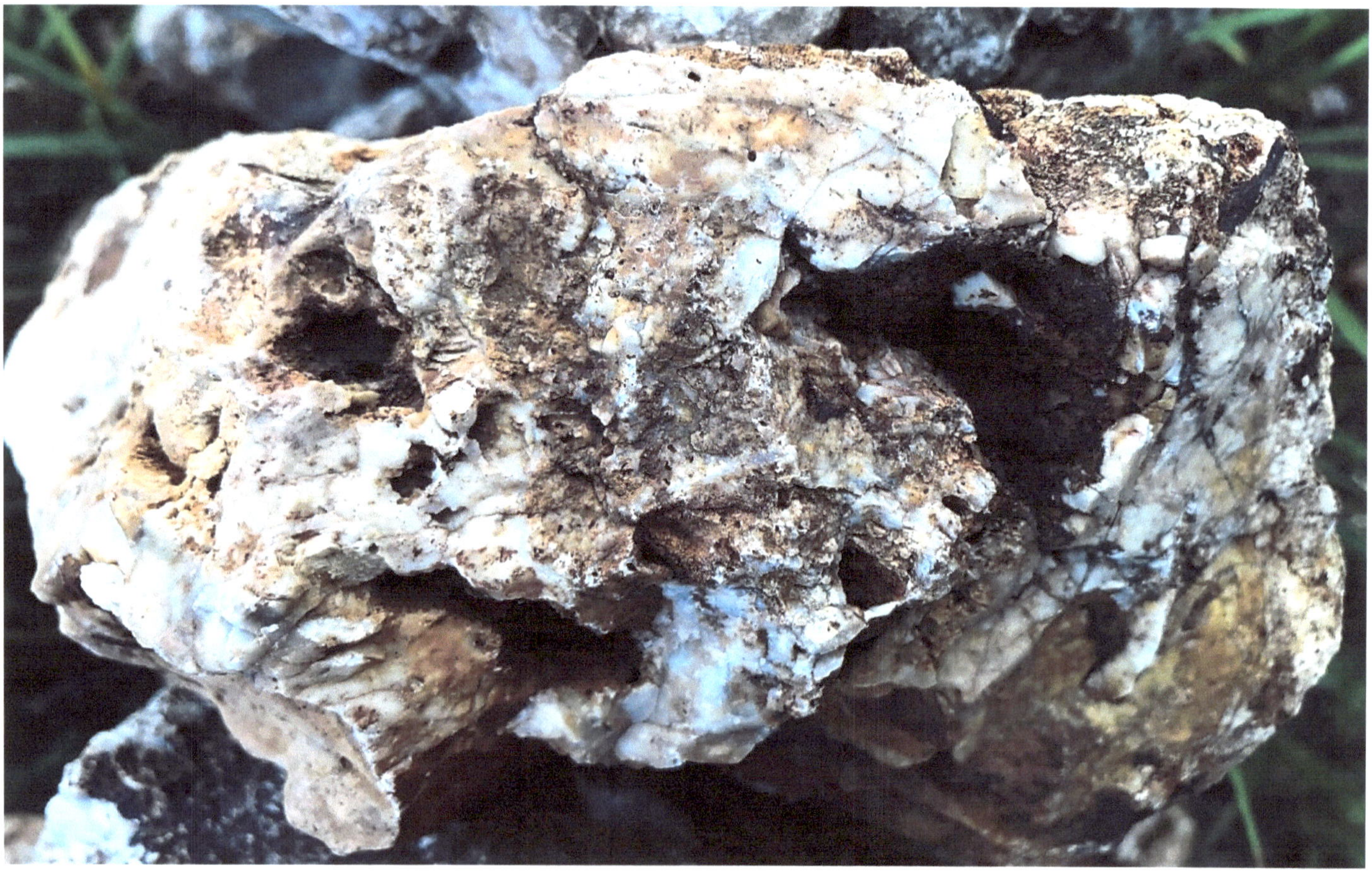

Horned Head Rock

Rock Guardian holds Horned Head Rock.

The Ocooch Mountain Rocks may look hard and durable, but many are fragile! We celebrated a holiday with a big bonfire, and two priceless rocks toppled into the fire and were destroyed. No replacements will ever be found. Each Ocooch Mountain Rock is a unique creation. Heat can cause rocks with less quartz to splinter into pieces. We are careful to have smaller fires and less heat!

Fossil stromatolite reefs can be seen worldwide, although they might not have rich colors or strange shapes. The Petrified Sea Garden in Lester Park near Saratoga Springs, New York, presents nearly an acre of fossilized stone, the first to be scientifically described.[69]

Known as the cave state, Athel Glyde Unklesbay, and Jerry D. Vineyard, in their book *Three Billion Years of Volcanoes, Seas, Sediments, and Erosion*, describe Missouri geology that features Ordovician sedimentary layers in the counties of Perry, Phelps, Pulaski, and Howell. Perry County has 630 known caves.[70]

If you go rock hunting, take time to turn over rocks lying on the ground or in piles. Something that looks ordinary might be a fantastic find! Power wash your rocks to remove dirt, grime, lichens, and moss and to reveal color. However, before you head for the hills, try Lake Michigan beaches. Glaciers scrubbed the area. Small, rounded stromatolite rocks can be found near the water. Pictures of them, coral stones, and agates can be seen on Pinterest.[71] I hope you enjoyed the book! May you find a rock you will always treasure!

B.C. the Cat

---

[69] Alles, ed., "Stromatolites," 27.
[70] Athel Glyde Unklesbay, and Jerry D. Vineyard, *Three Billion Years of Volcanoes, Seas, Sediments, and Erosion* (Columbia: University of Missouri Press, 1992), 52-53.
[71] "Lake Michigan Stromatolites Search," Pinterest, accessed December 15, 2020, https://in.pinterest.com/search/pins/?q=lake%20michigan%20stromatolites&rs=typed&term_meta[]=lake%7Ctyped&term_meta[]=michigan%7Ctyped&term_meta[]=stromatolites%7Ctyped.

Bibliography

Alles, David L. Ed. Stromatolites." Western Washington University PDF 1-29. Last updated March 28, 2012. https://fire.biol.wwu.edu/trent/alles/Stromatolites.pdf.

Allwood, Abigail C, Malcolm R. Walter, Balz S. Kamber, Craig P. Marshall, and Ian W. Burch. "Stromatolite reef from the Early Archaean era of Australia." *Nature,* 441 (July 2006): 714–718. DOI: .1038/nature04764.

Archibald, J. M. "Secondary Endosymbiosis." In Encyclopedia of Microbiology, 3rd ed Amsterdam: Elsevier, 2009, 438-446. https://www.sciencedirect.com/science/article/pii/B9780123739445003606.

Battersby, Stephen. "Earth's wild ride: Our voyage through the Milky Way." *NewScientist,* November 30, 2011. https://www.newscientist.com/article/mg21228411-500-earths-wild-ride-our-voyage-through-the-milky-way/.

Bekker, Andrey. "Huronian Glaciation." In *Encyclopedia of Astrobiology.* Berlin, Heidelberg: Springer, 2011, 768-772. https://doi.org/10.1007/978-3-642-11274-4_742.

Bourgeois, Joanne. ROBERT H. DOTT, JR., (1929–2018). EARTH HISTORIAN AND HISTORIAN OF GEOLOGY. *GeoScienceWorld* 38, 2 (October 1, 2019): 422-429. https://doi.org/10.17704/1944-6178-38.2.422.

Davis Jr., Richard A., Robert H. Dott, Jr., and Ian W. D. Dalziel. "Proterozoic geology of the Baraboo Interval." In *Geology of the Baraboo, Wisconsin, Area,* 43. Boulder: Geological Society of America, January 1, 2016, 7-12. https://doi.org/10.1130/2016.0043(05).

Djokic, Tara, Martin J. Van Kranendonk, Kathleen A. Campbell, Malcolm R. Walter & Colin R. Ward. "Earliest signs of life on land preserved in ca. 3.5 Ga hot spring deposits." *Nature Communications,* 8, 15263 (2017): 1-7. https://doi.org/10.1038/ncomms15263.

Dott, Robert H. Jr. and John W. Attig. *Roadside Geology of Wisconsin.* Missoula Montana: Mountain Press Publishing Co., 2004.

Fox, Douglas. "What Sparked the Cambrian explosion?" *Nature News Feature,* February 16, 2016. https://www.nature.com/news/what-sparked-the-cambrian-explosion-1.19379.

Ghose, Tia. "3.5-Billion-Year-Old Fossil Microbial Community Found." *Live Science,* November 13, 2013. https://www.livescience.com/41191-ancient-microbe-fossils-found.html.

Gower, Charles, and Paul Clifford. "The structural geometry and geological history of Archean rocks at Kenora, north-western Ontario— a proposed type area for the Kenoran Orogeny." *Canadian Journal of Earth Sciences,* 18 (2011): 1075-1091. https://doi.org/10.1139/e81-103.

Kumar, Anil, Ravi Shankar, Parashuramulu Vadlakonda, and J., Besse. "Evidence for a Neoarchean LIP in the Singhbhum craton, eastern India: implications to Vaalbara supercontinent." *Precambrian Research* 292 (2017): 163-174. https://doi.org/10.1016/j.precamres.2017.01.018.

Hazen, Robert M. "Evolution of Minerals." *Scientific American,* 302, 2 (2010), 58-65. https://doi.org/10.1139/e81-103.

International Commission on Stratigraphy. "Chart/Time Scale." Accessed July 15, 2019. https://web.archive.org/web/20170113013553/http://www.stratigraphy.org/index.php/ics-chart-timescale.

Long, Major Stephen. *Voyage in a six-oared skiff to the Falls of St. Anthony in 1817.* Philadelphia: Henry B. Ashmead, Book and Job Printer, 1860.

Mahapatro, S., Naresh Pant, Santanu Bhowmik, A. Tripathy, and Jayanta Nanda. "Archaean granulite facies metamorphism at the Singhbhum Craton-Eastern Ghats Mobile Belt interface: Implication for the Ur supercontinent assembly." *Geological Journal,* 47, 2-3 (2012): 312-333. Doi.org/10.1002/gj.1311.

Marshall, Jessica. "Geology: North America's broken heart." *Nature News Feature,* December 4, 2013. https://www.nature.com/news/geology-north-america-s-broken-heart-1.14281.

Martin, Lawrence. *The Physical Geography of Wisconsin.* Madison: University of Wisconsin Press. 1965.

Merriam Webster. "Karst." July 20, 2020. https://www.merriam-webster.com/dictionary/karst.

National Park Service. "Wisconsin." Accessed July 15, 2019. https://www.nps.gov/subjects/nnlandmarks/state.htm?State=WI.

Nutman, Allen P., Vickie C. Bennett, Clark R. L. Friend, Martin J. Van Kranendonk, & Allan R. Chivas. "Rapid emergence of life shown by discovery of 3,700-million-year-old microbial structures." *Nature* 537 (August, 2016): 535-538. https://doi.org/10.1038/nature19355.

Nuwer, Rachel. "What Happened to the Stromatolites, the Most Ancient Visible Lifeforms on Earth?" *Smithsonian Magazine Smart News,* May 30, 2013. https://www.smithsonianmag.com/smart-news/what-happened-to-the-stromatolites-the-most-ancient-visible-lifeforms-on-earth-84714880/.

Ojakangas, R. W., G. B. Morey, and D. L. Southwick. "Paleoproterozoic basin development and sedimentation in the Lake Superior region, North America." *Sedimentary Geology,* 141–142 (2001): 319-341. https://doi.org/10.1016/S0037-0738(01)00081-1.

Orr, Isaac. "Minnesota's Copper and Nickel Deposits are World Class." Americanexperiment.org. March 12, 2018. https://www.americanexperiment.org/2018/03/minnesotas-copper-nickel-deposits-world-class/.

Oskin, Becky. "Crash! 10 Biggest Impact Craters on Earth." *Live Science,* April 28, 2014. https://www.livescience.com/45126-biggest-impact-crater-earth-countdown.html.

Palacas, James G. "Superior Province (051)." USGS Energy Data Finder PDF. Accessed July 15, 2019. https://certmapper.cr.usgs.gov/data/noga95/prov51/text/prov51.pdf.

Peakery. "Mt. Whittlesey." Accessed July 15, 2019. https://peakery.com/mount-whittlesey-wisconsin.

Penn State. "First Land Plants And Fungi Changed Earth's Climate, Paving The Way For Explosive Evolution Of Land Animals, New Gene Study Suggests." *ScienceDaily,* August 10, 2001. www.sciencedaily.com/releases/2001/08/010810070021.htm.

Perkins, Sid. "Giant Asteroid Impact Could Have Stirred Entire Ocean." *Wired Science,* November 10, 2009. https://www.wired.com/2009/11/giant-asteroid-impact-could-have-stirred-entire-ocean/.

Pinterest. "Lake Michigan Stromatolites." Accessed December 15, 2020. https://in.pinterest.com/search/pins/?q=lake%20michigan%20stromatolites&rs=typed&term_meta[]=lake%7Ctyped&term_meta[]=michigan%7Ctyped&term_meta[]=stromatolites%7Ctyped.

Princeton University. "A more potent greenhouse gas than carbon dioxide, methane emissions will leap as Earth warms." *ScienceDaily*, March 27. 2014. https://www.sciencedaily.com/releases/2014/03/140327111724.htm.

Rizzo, Vincenzo and Nicola Cantasano. "Structural parallels between terrestrial microbialites and Martian sediments: are all cases of Pareidolia'?" *International Journal of Astrobiology,* 16, 4 (2016): 297-316. https://doi.org/10.1017/S1473550416000355.

Schirber, Michael. "Snowball Earth Might Have Been Slushy." NASA News Features. August 2015. https://www.giss.nasa.gov/research/features/201508_slushball/.

Schmus, W. R. and W. J. Hinze. "The Midcontinent Rift System." *Annual Review of Earth and Planetary Sciences* (May 13, 1985): 345-383. Doi:10.1146/annurev.ea.13.050185.00202.

Schultz, Gwen. *Wisconsin's Foundations* 2nd ed. Madison: University of Wisconsin Press, 2004.

Schultz, Klaus J., and William F. Cannon. "The Penokean orogeny in the Lake Superior region." *Precambrian Research* 157, 1–4 (2007): 4-25. https://doi.org/10.1016/j.precamres.2007.02.022.

Sharkbay. "Stromatolites." Accessed December 15, 2020. https://www.sharkbay.org/publications/fact-sheets-guides/stromatolites/.

Specktor, Brandon. "Earth's oldest known meteor crash site found in Australian Outback." *Live Science,* January 22, 2020. https://www.livescience.com/worlds-oldest-meteor-crater-yarrabubba.html.

Smith, George L., Robert H. Dott, Jr., and Steven W. Byers. Authigenic Silica Fabrics Associated with Cambro-Ordovician Unconformities in the Upper Midwest." *Geoscience Wisconsin*, 16 (1997): 25-36. https://wgnhs.wisc.edu/catalog/publication/000233/resource/gs16a03.

State Cartographer's Office, University of Wisconsin-Madison. "Wisconsin High Points." Updated August 14, 2017. https://www.sco.wisc.edu/wisconsin/high-points/.

St-Onge, Marc R., Michael P. Searle, and Natasha Wodicka. "Trans-Hudson Orogen of North America and Himalaya-Karakoram-Tibetan Orogen of Asia: Structural and thermal characteristics of the lower and upper plates." *Techtonics,* an AGU Journal, 25, 4 (July 18, 2006): PDF. https://doi.org/10.1029/2005TC001907.

Tollo, Richard, Louise Corriveau, James McLelland, and Mervin Bartholomew. "Proterozoic tectonic evolution of the Grenville orogen in North America: An introduction." *Memoir of the Geological Society of America,* 197. Boulder: Geological Society of America, January 1, 2004, 1-21.

Universitaet Tübingen. "Iron in primeval seas rusted by bacteria." *ScienceDaily,* April 23, 2013. https://www.sciencedaily.com/releases/2013/04/130423110750.htm.

University of California-Davis. "Oxygen oasis in Antarctic lake reflects Earth in distant past." *ScienceDaily,* September 1, 2015. https://www.sciencedaily.com/releases/2015/09/150901140759.htm.

Unklesbay, Athel Glyde, and Jerry D. Vineyard. *Three Billion Years of Volcanoes, Seas, Sediments, and Erosion.* Columbia: University of Missouri Press, 1992.

Vanchard, Daniel. "Cyanobacteria." In *Encyclopedia of Geology*, 2nd Ed. Amsterdam: Elsevier Ltd, 2021, 446-460. doi.org/10.1016/B978-0-12-409548-9.11843-3.

Wade, Grace, "1.75-billion-year-old fossils help explain how photosynthesis evolved," *New Scientist*, January 3, 2024, https://www.newscientist.com/article/2410391-1-75-billion-year-old-fossils-help-explain-how-photosynthesis-evolved/.

Weisberger, Mindy. "Slowdown of Earth's spin caused an oxygen surge." *Live Science*, August 2, 2021. https://www.livescience.com/early-earth-rotation-increase-oxygen.html.

Worldatlas. "What Is A Supercontinent And A Superocean?" August 16. 2017. https://www.worldatlas.com/articles/what-is-a-supercontinent-and-a-superocean.html.

Zimmer, Carl. "Earth's Oxygen: A Mystery Easy to Take for Granted." *The New York Times,* Oct. 3, 2013. https://www.nytimes.com/2013/10/03/science/earths-oxygen-a-mystery-easy-to-take-for-granted.html.

Index

algae, cyanobacteria entered cells with DNA, 17
Algoman Orogeny, mountain building event, 14
Animikie Sea, shallow epi-continental sea, 15-16
Authigenic, formed in place, not transported, 5
Baraboo Hills, formed 1.7 billion years ago, 16
Barlangi Hill, impact site, ended Huronian glaciation? 15
carbonic acid, creates cave cavities, 7-8
carnivory, additional oxygen a factor, 18
Chamberlin
    T. C., UW pioneering geologist, 3
    described Ocooch Mountain Rocks, 3
    explained why Driftless Area is not glaciated, 3
Laurentia, early North America, 13
Cyanobacteria
    associated with
        banded iron deposits, 15-16
        stromatolite stone formations, 18
        the Great Oxygen Event, 15
    fossils dated
        3.5 billion years ago in Pilbara Craton, 13
        3.7-3.8 billion years BP in SW Greenland, 14
        1.75 billion years BP with chlorophyll, 14
    photosynthetic bacteria, 13
    survived Snowball Earth, 15, 17
    traced through geologic history, ix
diagenesis, chemical alteration of rock, 3–5, 8
disappearing streams, signs of a cave, 18
dolomite
    capstone of Driftless Area peaks, 7
    interacts with water to form carbonic acid, 7
    limestone w/magnesium, vital mineral for life, 3, 7-8
Driftless Area
    southwestern Wisconsin, 6
    unglaciated, 6
Ediacaran Period, early life proliferated in, 18
Gogebic, Wisconsin banded iron range, 16
Great Lake Tectonic Zone, almost an ocean, 16
Great Oxygen Event
    2.5 billion years ago, essential for life, 14
    longer days may have caused, 15
Grenville Orogeny, sutured GLTZ 17
hydroxides, create color in rock, 6
iron
    band accumulation, 16
    oxides create color, 13
karst, topography, from kräs, 18
Kenoran Orogeny, 14, 86 *See also Trans-Hudson*
Kenorland, early continent, Great Oxygen Event, 14
Lake Michigan, stromatolite fossils near shore, 87
Laurentia, early continent, 13
Lerova
    early ocean, 14
    monsoons, 14
lichens, 17
Long, Major Stephen, first to describe Ocooch Mtn. Rocks, 5
Mesabi, Minnesota banded iron range, 16
Ocooch Mountains, in the Driftless Area, 6
Oneota, sandstone, stromatolites found in, 3
Oolite, rock fragments ground to ball-bearing shape, 4
Ordovician
    lower, has stromatolites in the Driftless Area, 3
    stromatolites proliferated in shallow sea, 4
    sea withdrew, rivers cut deep into stromatolite stone, 3
pareidolia, the tendency to see patterns, 19
Penokean Mountains
    high as the Alps, 16
    numerous water gaps, 16
Prairie Chien Dolomite
    associated with the Ocooch Mountain Rocks, 3, 7-8
    Oneota, lower Ordovician, has stromatolite fossils, 3
river gravel, found in an ancient river bed, 6
Rodinia, supercontinent, 17
silicification
    replacement of dolomite by silica, quartz, 6-8
    sinkholes, signs of a cave, 18
snowball earth
    Huronian, 15
    Cryogenic Age, 17
St. Peter Sandstone, non-marine, covered Oneota, 3, 8
stalactites, form on cave ceilings, 7
stalagmites, form on cave floors, 7
stromatolite
    associated with Prairie du Chien dolomite, 3, 7-8
    cyanobacteria structures, mucus traps sediment, 4
    formed in tropical offshore shoals, 4
    from the Greek word strōma (layer), 3
    found
        in Oneota sandstone, lower Ordovician Period, 4
        at Shark Bay, Lake Thetis, Australia, 4-5, 26
        on Mars?, 14
    small biscuit to large turban shapes, 4
    structures chemically altered by diagenesis, 3
    stacked, 7
    stems form in deep water, 4
    tops round in sunlight, 4
    types
        hemispherical, raised circular, biscuit, crayback,
        microbiotic towers, 10, conophytes, 25
Stromatolite Images and Gallery
    Animal-Shape, 20
    April's Owl and Magnified Quartz, 56-61
    Bird on a Stone Nest, 19
    Bird-Shape, 20
    botryoidal, 5, 24
    bowl-like fragments, 19
    breccia, fused angular rock fragments, 3, 8
    cave fragments, 7
    Circular Dome, 11
    Christmas Tree, 24
    Dinosaur Skull and Stone Face, 23
    Dragon with white fire, 31
    Drusy grotto, 3
    Eye of the World, 84
    Figurine, 20
    Garden, 2, 4
    Ghost Rock, 22
    Ghost Rocks in the Snow, 23
    Horseshoe, 19
    impressions, 19
    intact, 5
    iron crust, 12
    Jurassic Park Creature, 30

Scary Rock Fish, 31
Mystical Rock, 24
Odin Owl, 32
Oracle Rock, 20
Old Stromatolite, 25
Old Woman with Baby, 50
Owl Family in a Stone Tree, 21
Prairie Dog Pups, 20
quartz
    amethyst, mega, 5-6
    black spots, 28
    smooth, 28
    red, 5
    ribbon, 27
    round, 28
Rock
    Creature, 36
    Dragon, 53
    Eyes, 66-73
    Faces
        Collage. 62
        Strange, 63
        Ghostly, 54-55
        in Stone Blocks, 65
        Stone, 22
    Guardian, 83
    Head, 37-49
    Horned, 83
    Mother's, 1
    Snakes, 34-35
    Spiders, 33
    Troll, 20
    Skull, 79-82
    Stare, 74-78
    Voices, 64
    Ropy Convolutions, 9
    with hole the remaining feature, 24, 26
    sedimentary lines, 3-5
    stem, 3, 22
Stone Creature, 20
Stone Eye, 27
Stone Garden Creature, 30
Stone Horse with many horns! 52
Stone with small cavities, 4
types
    Hemispherical, oncolite, raised, circular, crayback, microbialite towers, 10, conophyton, 25
vuggy, 5, 12
Woman with long hair, 51
Sudbury impact, ended iron band form. in Animikie Sea, 16
Supercontinent Cycle, continents break up & merge again, 14
Superior Province, one of the earliest cratons, 14
Trans-Hudson Orogeny, high as Himalayas, 16
Thrombolites, cyanobacteria mats infected by foraminifera, 18
Ur, second oldest hypothetical continent, 14
Vaalbara
    earliest hypothesized continent, South Africa's Kaapvaal craton (Vaal) joined with Australia's Pilbara (ara), 14
Wisconsin Dome, erosion exposed lower Ordovician, 6
Yilgarn Craton, one of the first cratons, Australia 13
unconformities, break in sedimentary rock, 5–6, 7-8

www.ingramcontent.com/pod-product-compliance
Lightning Source LLC
Chambersburg PA
CBHW041529220426

43671CB00002B/28